THE AUTOBIOGRAPHY OF A TRANSGENDER SCIENTIST

THE AUTOBIOGRAPHY OF A TRANSGENDER SCIENTIST

BEN BARRES

The MIT Press
Cambridge, Massachusetts
London, England

The text of this book is based on Ben Barres's chapter in *The History of Neuroscience in Autobiography* published by the Society for Neuroscience.

This book was set in Scala by Scribe Inc. Printed and bound in the United States of America.

Library of Congress Cataloging-in-Publication Data

Names: Barres, Ben, author.
Title: The autobiography of a transgender scientist / Ben Barres ; foreword by Nancy Hopkins.
Description: Cambridge, MA : The MIT Press, [2018] | Includes bibliographical references and index.
Identifiers: LCCN 2018011624 | ISBN 9780262039116 (hardcover : alk. paper)
Subjects: LCSH: Barres, Ben. | Neurobiologists—United States—Biography. | Transgender people—United States—Biography.
Classification: LCC QP353.4.B37 A3 2018 | DDC 612.8092 [B]—dc23 LC record available at https://lccn.loc.gov/2018011624

10 9 8 7 6 5 4 3 2 1

CONTENTS

FOREWORD

Many people tried to convince Ben Barres to write a book about his extraordinary life. He didn't have time. Then on March 30, 2016, came his heartbreaking e-mail:

Subject: Some sad news

Dear Nancy,
Alas, I have just been diagnosed with advanced metastatic pancreatic cancer. My prognosis is probably only a few months at best. That said, I have a brca2 mutation that predisposed me to this and a subset of those sometimes live a year or more on chemo that includes parp inhibitors, which I will receive. . . . I am in the Stanford hospital as the tumor . . . caused me to have a heart attack. All and all I am extremely ill.

Ben died on December 27, 2017. During the twenty-one months after his diagnosis he found time to write this book. He also continued to run his lab, publish scientific papers, mentor his trainees, and ensure that the letters of recommendation they would need to advance their careers after he died were up to date. He traveled to Hawaii to lecture, continued his advocacy work for women and minorities in STEM and for the LGBT community, attended an all-day scientific symposium where over three hundred of his trainees and colleagues gathered to celebrate his life, took his students to one more Harry Potter movie, and bicycled through the Palo Alto hills between rounds of chemotherapy and experimental cancer treatments. In addition, he maintained his extensive e-mail correspondence. I stopped deleting any of his e-mails when I learned he was ill. I have 496 on my computer, some of which I share below in keeping with Ben's spirit of utter candor and openness, and to capture his unique voice.

The outline of Ben's story is widely known: Born female, it was Barbara Barres who fell in love with science as a child, matriculated at MIT, got an MD at Dartmouth and a PhD at Harvard Medical School, then went to London for postdoctoral training. Highly successful at every stage of her education and training, it was Barbara who in 1993 accepted a faculty job in Stanford's neurobiology department. But at age forty-three, Barbara transitioned to male and became Ben. Later he became department chair of neurobiology at Stanford.

An uncontrollable passion for science drove Ben. He was renowned for his lab's discoveries in neuroscience. More than half of human brain cells are a type of cell called glia, but glia were long neglected by scientists in favor of the more glamorous neuron—the cells that transmit the brain's messages. Ben thought it impossible that glia were not extremely important to brain function, so he decided to study them. Over decades, he discovered critically important roles for glia, securing his scientific legacy and launching the careers of scores of young scientists who now advance the field he and they essentially created.

Had Ben been a less spectacular scientist he would still have had a major impact on medical research as a mentor and advocate for young scientists. Elite science is hyper-competitive, and individual faculty wield enormous power and influence over their trainees. Doctoral students and postdocs can get ground up by professors more intent on building their own CVs than the careers of their trainees. Ben was merciless in his criticism of such behavior. For him scientific research was a privilege and a joy to be shared. His trainees became family—his children; their success was more important to him than his own. He educated, and when needed, chastised and prodded the scientific community to be better teachers, more generous and caring mentors. He wrote two widely read articles on the subject—one advising students how to find a mentor, the other (written when he knew he was dying) urging faculty not to compete with their own students but to be generous and let young scientists take their discoveries with them to launch their independent careers.

But, remarkably, Ben was more widely known not for his magnificent science, or for elevating mentoring to an art form, but for his advocacy on behalf of women in science. That is how we became friends, since I too am a biologist who advocated on behalf of women.

To understand the impact of Ben's advocacy requires a brief history of women's advancement in academic science in the Barbara/Ben era. To explain how Ben single-handedly helped to level the playing field requires examples, including examples of Ben's inimitable style of persuasion. First, some history.

When Barbara Barres went to college in 1972, as with young women today, it didn't occur to her that she would encounter gender discrimination. The 1964 Civil Rights Act and affirmative action efforts and women's liberation in the late 1960s and early 1970s were taking care of that problem. Besides, science is supposed to be a meritocracy and Barbara was a brilliant student. True, she had occasional inexplicable experiences: As an MIT undergrad, Barbara was the only student in a class to solve an unusually challenging math problem but the professor refused to give her credit. He accused her of cheating, saying her boyfriend must have solved it.

But Barbara didn't have a boyfriend. To her, the professor's behavior was incomprehensible (and intolerable) unfairness, not obviously gender discrimination. As for the professor, he knew Barbara Barres *must* have cheated. What other explanation could there be since, as everyone knew back then, "girls can't do math."

Women who had started their training and scientific careers a decade earlier than Barbara, had begun to make sense of these un-fairnesses. They came to understand that they constituted a previously unrecognized form of discrimination that today we call unconscious gender bias. Psychologists had already discovered this bias by controlled experiments but most women scientists discovered it through personal experience. Their discoveries began to emerge in the 1990s and shocked the scientific community. In 1997 two Swedish women scientists sued to obtain the records of a prestigious postdoc fellowship competition and discovered that a woman had to be 2.5 times as productive as a man to win a fellowship. In 1999 the president of MIT endorsed a report from a committee I had chaired that documented that tenured women scientists were paid less, given fewer resources, and received fewer awards than men of comparable or even lesser accomplishment. Many universities soon reported similar findings. And a 1998 book, *Why So Slow*, by Virginia Valian, summarized research by psychologists demonstrating that a woman's work is valued less than identical work done by a man, providing a convincing explanation for women's experiences.

And what about Barbara Barres? By the 1990s she too had accumulated her inexplicable experiences of un-fairness. Soon she would have irrefutable evidence that they constituted gender bias. Having been a highly successful woman scientist, Barbara transitioned to Ben and learned as few of us ever could that scientific merit depends on whether people think you are a man or a woman. Not long after her transition Ben gave a seminar after which an attendee was overheard to say, *"Ben Barres gave a great seminar today, but then his work is much better than his sister's."* A radical activist for women in STEM was born.

About to cross paths with Ben the activist was Harvard president Larry Summers. In 2005, speaking at a conference convened by experts to discuss remedies for the under-representation of women in STEM, Summers hypothesized that unconscious bias was unlikely to be a significant cause of women's under-representation; rather, he said, women choose family over career, and furthermore, women may lack the intrinsic ability to populate the ranks of Harvard's science faculty: in short, they are genetically inferior. Known for being a provocative contrarian, Summers chose to overlook the breakthrough discoveries of the 1990s. Instead, he fell for the oldest, still least scientifically supported explanation for the different status of groups of people within a society: their genes.

Shock waves reverberated among professional women scientists and engineers who had been battling this belief and its devastating fallout for decades. The Summers speech was front page news for months. Women knew that if even a few powerful scientists came forward to say Summers was right, it could cost women their fragile progress. Much needed was a scientist with impeccable credentials to explain to a wide audience that there's not a shred of credible evidence to support the belief that women can't excel at high-level math and science in numbers comparable to men.

In 2006, Ben published the article "Does Gender Matter?" in the journal *Nature*. Using the power of his personal story, his knowledge of the psychology research, a lifetime of studying the brain, and a full awareness of freedom of speech issues and academic freedom, Ben asked in withering prose, "If a president of a prestigious university is going to pronounce in public that women are likely to be innately inferior, *would it be too much to ask that they be aware of the relevant data?*"

The article is an intellectual tour de force and a landmark on the path to women's equality, defining where we were at the turn of the twenty-first century. It went viral—covered by major newspapers, TV news programs, and journals around the world. It proved to be just the start of Ben's career as an activist. In the same article Ben proposed ways in which scientific institutions needed to change to reduce the impact of gender bias on women's careers. He became a one-person guerilla-army that achieved many of the goals he had outlined. He successfully:

- Changed the procedures used by NIH to select winners of the prestigious Pioneer Award (after an initial nearly-all-male committee selected almost exclusively male winners).
- Changed the process for nominating candidates to be chosen as Howard Hughes Medical Institute (HHMI) investigators (which previously had inadvertently resulted in women almost never being nominated).
- Established the child care assistance program for untenured faculty in conjunction with Stanford's provost.
- Convinced (some) scientific conferences to require a pledge from attendees that they will not sexually harass other attendees at meetings, with the understanding that, if they do, they will be banned from further participation.

To hone his skills, Ben took on individuals who transgressed: university presidents who had failed to control the bad behavior of faculty; conference organizers and faculty responsible for inviting seminar speakers who didn't invite enough women to speak; institutions that failed to display portraits of noted female scientists alongside those of male scientists. All received clear instructions on how to change their ways.

How does one person convince institutions to change their policies, individuals to change their behavior? Ben's scientific clout, the fact that he so clearly did not benefit from the causes he espoused but rather risked damaging his own career by doing so, and of course the fact that he was on the "right side" of the argument all contributed. But his personal style and his persistence were essential, as some examples Ben sent me show:

Dear Sir,
You have a hell of a lot of nerve inviting me after sending me that speaker list.

it looks like out of your last 35 speakers, only 1 has been a woman??!

I wouldn't visit your school if you were the last school on earth. Do you think women are not doing equally good science? And what about the half of your trainees that are women (not to mention the men)? Do you really mean to teach them that the only ones worthy of inviting are men??

I would suggest if you want to hear a really great talk about glia that you invite some of my previous women trainees to come speak in my place . . . [here Ben names three women]. Any of them would give a tremendously good talk. Looks like your faculty is not particularly diverse either (1 woman out of 17?????!!!!). I always thought Netherlands was one of the good places. Looks like I was wrong.

I am going to decline your invitation but perhaps in 5–10 years if you have cleaned up your act. I won't be holding my breath. Something is really rotten about your department.

Please don't invite me again

Ben

While such a missile might be met with silence or anger, when it came from Ben the reply was anything but:

You have every right to be upset about this very major issue. In fact, the Netherlands has one of the very worst records among Western countries of female university professors and business executives. Shortly after my arrival to the Netherlands, I specifically discussed this issue with the Dean of our medical school. There is indeed a long way to go. . . . I deeply respect and fully agree with your objection . . . please be sure that I absolutely will discuss your reply with the faculty of the department and the university administration, so that they understand the inexcusable standard which now exists and the severity of the consequences. Moreover, you have my promise to redouble my own efforts to improve the representation of women in science. Kind regards,

In 2015 Ben was particularly upset over sexual harassment at scientific meetings and began to ask meeting organizers to put policies in place to prevent such behavior. He also realized it would be more efficient if such policies were mandatory for NIH-supported meetings and wrote to the Director of the NIH. Here is a classic example from Ben's years-long exchange with the NIH director:

January 16, 2016
Dear Dr. Collins,
With respect, I have repeatedly written to you over the last year or two to ask that NIH take a simple step to protect women trainees

from sexual harassment at meetings. You never respond to me and nothing appears to be done and I cannot help but be deeply angry with you. I simply ask that before NIH funds a biomedical meeting (Gordon Conference, Keystone, Cold Spring Harbor, etc) that the meeting center have a sexual harassment policy in place. All the meeting hall would have to do is ask each meeting attendee when they register for the meeting to click on acceptance of this policy (which would say that faculty hitting on trainees for sex is not permissible and that bad behavior will be addressed by excluding attendee from future meetings).

Where is your decency? Have you talked to any women trainees? They CANNOT attend these meeting without being "hit on" continuously by male faculty. I told you previously of one famous 60 year old faculty member who brags of bedding over 200 trainees at meetings in his career. Another trainee told me recently that she was having what she thought was a terrific science conversation at her poster with a senior scientist who at the end of the conversation wrote his hotel room number on her hand.

I DO NOT WANT MY TRAINEES SUBJECT TO THIS BEHAVIOR ANY MORE. What is the big deal about simply asking for this simple thing on these meeting grants? It is trivial for NIH to do and trivial for the meeting halls to do and it would finally protect our trainees. Without this protection, women trainees are very effectively being prevented from networking at meetings and continuously being given the message that they are only valued for sex. Please lets put an end to this. I know that you are a very decent man and that you care about stuff like this, but I also know that you are very busy—please elevate this to a higher level of your attention. It really is critically important.

This is the last time I am going to write you. If nothing is done, the next letter I will be writing will be to Senators Kirsten Gillibrand and Claire McCaskill.

Thank you and sincerely,
Ben A. Barres, MD, PhD
Professor and Chair of Neurobiology
Stanford University

A few days later Ben received a gracious reply from Collins, thanking him for keeping the issue "front and center" and detailing NIH's latest efforts to address the problem. Collins concluded by seeking Ben's "advice on any additional practical steps that NIH, as a research funding agency, could take." Not long after this exchange, I read in *Nature* magazine (3 March 2016) about an "NIH push to stop sexual harassment." I wrote to congratulate Ben but he was not at all satisfied; in fact he was upset. The article suggested that further study was needed while Ben wanted action, not studies. Needless to say, Ben took Collins up on his request to provide "additional advice" and the correspondence continued. Persistence was a hallmark of Ben's approach.

I asked Ben where he found the courage for this type of advocacy when so many people are reluctant to speak out.

its not particularly stressful for me nancy
maybe its because i am transgender
i really don't care what other people think of me
and i really couldn't care less who i piss off
thats the virtue of tenure isn't it?!
whats stressful to me is having such an unfair world

Given how effectively Ben could lead from his position of department chair I wondered what he might do if he were director of NIH himself, or president of a university. One day I saw Ben's name on a list of people recommended to be president of the Rockefeller University and wrote to tell him I thought he'd be terrific. But he replied:

no one on that list is ready to give up their research to be an administrator! definitely not me! and i am the last person that would make a good president.

if they made me president of Stanford, my first two acts would be to blow up the campus church (along with banning priests and prayers from graduation events) and to get rid of stanford football!

at least rockefeller doesn't have either of those!

Which brings me to something deeply mysterious about Ben that may be impossible to explain. For all his outrage at the world's unfairness, the pain that accompanied his early gender confusion, the risks he took and people he might have offended, he was the most beloved scientist I've known, and the most loving. He could charm you with a conspiratorial smile as he contemplated some new unfairness he planned to tackle using "the nuclear option." Or envelop you in love and support when he sensed you were flagging or fearful. One felt safer because he existed. When I knew Ben was close to dying I wanted to make sure he knew how much I loved him, although it was hard to know what to say.

Dear Ben,
I was trying to figure out yesterday why so many people love you. Admire I understand. But that people love you so much. What's that about? Don't like to bother you with fan mail, but I miss you.

Nancy

He replied,

> Thank you so much Nancy. You know it's impossible for me to imagine anyone loving me, or even liking me. Too much emotional scarring left over from an LGBT childhood I guess.
>
> I am home on hospice now. . . . My previous and current trainees have been spending a lot of time visiting me which has been wonderful.
>
> And thank you for the fan mail; I don't get it very often!

The obituaries and tributes for Ben were a magnificent collection of fan mail as people struggled to express their loss and to capture the essence of this extraordinary person who improved the world in so many ways. And I still wonder in years to come how far his legacy will extend. For now, a comment from his friend Jo Handelsman sums it up well: "Ben: You leave a towering legacy of goodness. Thank you."

Nancy Hopkins
MIT

INTRODUCTION

I have not yet retired but was diagnosed about two years ago, at the age of sixty-one, with advanced pancreatic cancer. Although this is generally considered one of the most aggressive and least treatable of cancers, thanks to recent medical advances, I am fortunate to still be working every day. Surprisingly, cancer has had an upside. It allowed me to shed many time-consuming activities such as being department chair, being on endless committees and editorial boards, grant writing, teaching, and traveling to meetings. It suddenly became easy to say no to all these things without guilt so that I could focus on the things I really wanted to accomplish before my time runs out.

What was left was all the very best parts of being a scientist, running a lab and mentoring young scientists. I am grateful for this opportunity to write about my life. I would like to tell you what a privilege it has been to be a scientist and to mentor young scientists. I would also like to tell you about my experiences as a female scientist, then as a transgender scientist, and how my differences may have contributed beneficially to my path in science. Finally I would like to tell you about glial cells, what we know as well as what we don't know, and the great adventure it has been to explore their roles in healthy and diseased brains.

LIFE

GROWING UP

I was raised in West Orange, New Jersey. My family was not financially well off. There were four of us kids, so my mom spent her time raising us, while my dad worked as a salesman, first of baby furniture and later of liquor. My mother came from a Lithuanian Jewish family and my dad came from an Italian Roman Catholic family. Neither of my parents attended college but my mom was highly intelligent and she expected that her kids would do well in school and attend college.

As a woman, she had been forced in high school to take the "secretarial" track. But when all of her kids had grown up, she enrolled at a local college, Rutgers, and started working toward a mathematics degree. Sadly in only her mid-forties she passed away from breast cancer, a familial curse resulting from a BRCA2 mutation, which I also inherited. She never lived to see how her kids did as adults, but she would have been proud. My fraternal twin sister Jeanne became a pediatrics nurse practitioner, my brother started a successful insurance company, I (Barbara) became a scientist, and my younger sister became a legal secretary. All of my siblings married and had children; I was the outlier in that regard, as I will come to.

My mom was not particularly compatible with my father, but they stuck it out for the good of us kids. Although there were many times when my parents struggled financially to make ends meet, we never lacked for basic necessities. Looking back on it, my mom never seemed very happy and she was often irritable. I suspect this was due to both undiagnosed severe obstructive sleep apnea—another familial curse—as well as the constant absence of my father. When he was not working, which was most of the time, he was either with his "second wife" (whom we only learned about after my mom died) or was out with his friends gambling, playing cards, or betting on ponies. From my mother, I inherited intelligence, the BRCA2 mutation, and sleep apnea, and from my father an addictive tendency, not for gambling but for doing research.

My parents apparently agreed when they married that their kids would be raised in the Roman Catholic religion and we only learned after my mom died that she had been Jewish. I do not know why she did not tell us this but undoubtedly she wished to shield her kids from anti-Semitism. So every Monday after school we would be bused to a local church for religious indoctrination and forced to attend mass at church every Sunday. Even in grade school I recognized that what I was being taught about God was not supported by evidence, was internally inconsistent, and made no sense. But I dared not upset my mom, who insisted that knowledge of right and wrong could only come from religion, an idea that deeply offended me then and now.

When I was about fourteen years old, I finally got up the courage to refuse to go to church anymore and informed my parents that I was an atheist. In response, my mother ordered my dad to beat me, which he did (this was the only time that he did this), and I ran away from home for a brief period. When I returned home, my mom informed me that until I started going to church, I would have to stay in my bedroom all day every Sunday without food. Needless to say every Saturday I would stock up my room with food and things to do. Fortunately, after a few months my mom relented and life resumed as normal.

I agree with those who have argued that it is a great crime to indoctrinate children with religious beliefs. It is always surprising to me that childhood religious indoctrination seems to stick lifelong for most people. For many kids, this indoctrination must exert a very powerful influence on the developing brain. Children not only quickly absorb knowledge but somehow must learn to neglect or not see the internal inconsistencies between actual facts and the fictions they are being taught. I wonder if such childhood indoctrination, by irreversibly affecting brain development, might have permanent effects on cognitive development, perhaps even impairing scientific thinking ability in later life.

One of the great pleasures of growing up was having endless time to read. My mom would take us to the West Orange Public Library once a week and we were allowed to take out two books, a limit that constantly annoyed me. Science was an early interest. At the age of four or five, my twin sister decided she wanted to be a nurse and I decided I wanted to be a scientist, and that is what we did. I do not know why I was interested in science at such a young age as I did not know any scientists and hardly had any concept of what science was. But somehow I had the idea that science was something fun. Perhaps it was because I liked to watch the show *Superman* on TV, which had a mad scientist who was always making fascinating concoctions and inventions. Perhaps that is why I thought I might someday be a chemist. My favorite childhood toys were chemistry sets and microscopes.

Unfortunately, from grade school onward, it always seemed to me that public school moved at a very slow pace. Fortunately I had a marvelous science teacher in eighth grade, Jeffrey Davis. He was full of passion for teaching and was one of the best teachers I had in public school. He made science discovery seem like incredible fun and I wanted to be in on it. I recall that we read the *Voyage of the Beagle* by Charles Darwin and also *The Double Helix* by James Watson, a book I have since read every ten years throughout my life. No book better captures the excitement of scientific discovery—and the realities that talented women face in science.

Starting in junior high school, I found an endless stream of local science programs at nearby universities to supplement my public school education. I was one of the best science and math students in my junior and high school classes and was captain of the math team. I attended mathematical astronomy courses at Rutgers University on weekends and during summers in junior high school. I was very fortunate to attend the Phillips Andover Academy summer session when I was about fourteen. They provided me with a full scholarship, as otherwise I would not have been able to attend; in recent years I have funded full scholarships for several Andover summer students with financial need. At Andover, I studied chemistry and calculus as well as computer programming. The six kids in the math class covered the entire calculus textbook in the six-week summer program, even though the class was only one hour per day.

At Andover I was able for the first time to take science and math courses that progressed at a more stimulating and challenging rate. When I returned home after the summer session was over, I keenly felt the loss as I faced returning to my local public school. Fortunately I learned of an NSF program at Columbia University called the Science Honors Program where Columbia faculty volunteered their time to teach high school kids science and math courses on Saturday mornings. Luckily I tested in and there was no charge to attend. For all of high school, every Saturday morning I would bus in to New York City to attend these courses. I am pleased that forty years later this wonderful program still exists. In these courses I found that many of my classmates came from the outstanding public high schools in New York City that specialized in science education, such as Stuyvesant High School and the Bronx High School of Science. I sure was jealous of these classmates who were able to attend such outstanding public schools full time.

Attending the Science Honors Program at Columbia allowed me to more fully develop my interest in computer programming that began at Andover. I spent much of the rest of high school learning computer languages (such as Basic, Fortran, assembly language, and C) and coding, which greatly helped lessen the boredom of high school. Laptop computers did not exist yet, so I had to find local universities that would allow me access to their computers, such as Columbia, Brooklyn Polytech, and Stevens Institute of Technology. This was great fun and prepared me for a wonderful opportunity at Bell Laboratories nearby in Murray Hill, New Jersey.

I was looking for a summer job in high school, and Bell Laboratories' computer systems division had an opening for a summer student. At my interview I was asked to spontaneously write short computer programs to do various tasks on the blackboard, which was a snap given all my computer experience, and luckily I won the job. My high school let me graduate a month early, so I was able to work the entire summer after high school as well as every summer during college. The salary they paid me greatly helped me afford to go to college, and they even provided me with a desperately needed full scholarship that financed my senior year in college.

Working at Bell Laboratories at that time was an incredible experience; they had not yet divested from the phone company and so the environment was still highly research intensive. I got to interact with a terrific group of engineers and computer scientists. Unix and C were just being invented in the department where I worked (my first summer I was coding in a language called "B"!). The computer skills I learned were to come in very handy when I attended college at MIT and later as a graduate student.

At Bell Labs I got to work in a research environment for the first time. From the start I found that once I was working on a project, I was totally hooked. I would race to work in the morning, stay late into the evening, and work Saturdays and Sundays as well. I would even debug computer programs while I was sleeping. This kind of self-motivated internal intensity has also characterized the research that I was to do as a PhD student, as a postdoc, and then in my own lab at Stanford.

In thinking about my success as a scientist, I do not attribute it to any especially great intelligence. I have met many people far more intelligent than I am who have been much less successful. I believe that two specific attributes have made me successful. First is the intense and uncontrollable passion that I have for doing research. I do not know where this passion comes from, but it has always been there. The second attribute is what in recent years has been called "grit" and refers to attributes of perseverance and resilience. I believe that I have grit in spades. I strongly suspect that this grit likely comes from my "difference," which I have omitted discussing up to now.

GENDER CONFUSION WHILE GROWING UP

I must admit that, despite boredom in public school, I had a fun time growing up. But there was a problem that I first became aware of at about the age of three or four years old. Although I was a girl, internally I felt strongly that I was a boy. This was evident in everything about my behavior. Because I had a fraternal twin sister, these differences were all the more apparent.

Starting in grade school, my sister had many girlfriends and liked to play with traditional toys associated with girls, whereas I strongly preferred to play with boys and toys that were more traditionally masculine. At Christmas, I was disappointed with gifts of jewelry or dresses and was always jealous of the toys that my brother, who was two years younger, would get. I desperately wanted to be in Cub Scouts and Boy Scouts, like my brother, but I got stuck in Brownies and Girl Scouts instead. Every Halloween, I would dress as an army man or a football player.

Unfortunately, as I got older, these differences were less and less acceptable. When I was about eight years old, I recall going over to my friend Tommy's house to play with his train set. But one day when I knocked on his door, his mother appeared and disapprovingly told me that I couldn't play with Tommy anymore because I should be playing with girls. From that time on I had few friends and learned to keep myself occupied with reading and various hobbies.

In junior high school, there were some new frustrations. I wanted to take wood shop, machine shop, and auto mechanics, but only the boys were allowed to take these classes, while the girls took sewing and cooking. Every year I asked permission to take the boys' courses and every year I was told no. But one year, one of the boys asked if he could take cooking because he wanted to learn how to make cookies. He was told yes!

In high school, my difficulties began to magnify. As I went through puberty, I felt uncomfortable with developing breasts, which I did not think I should have. And as my twin sister started to shave her legs and wear jewelry and make-up, I found all of this to be very uncomfortable. Instead, I dressed and acted as a tomboy. Whereas my sister had an active social life and many dates, I never dated in high school. Not that this bothered me, as I found that I had very little if any attraction to men (or to women). It was only much later as an adult that I finally realized that I lack the ability to experience sexual attraction (I also have severe face blindness and wonder if the two difficulties are connected somehow).

Because of my gender differences, I was often made fun of and bullied in high school. Another confusing thing was that, unlike my sisters, I never menstruated. Unfortunately, when I was about fifteen I was seen by a doctor who failed to examine me and injected me with high-dose estrogen monthly for over a year. It made me very ill but did not help. When it came time to go to college, I refused to take it anymore. I finally learned that, except for ovaries, I was born without inner reproductive organs, a condition known as Mullerian agenesis. This was another strong blow to my self-esteem, although lack of reproductive ability bothered me surprisingly little as I have never had maternal instincts or envisioned raising children.

From junior high school on, I had increasingly strong feelings of gender dysphoria, difference, and confusion. I felt very embarrassed and ashamed about my gender incongruity, but was totally unable to express what I was feeling to anyone. A male high school teacher once pulled me aside to lecture me about my tomboyish dress. I listened politely, but was embarrassed and unable to respond in any way. I never uttered a word about my gender confusion to my mother or to my siblings or to anyone else. Surely my mother must have noticed my unusual behavior in grade school, junior high school, and high school, but she never said a word about it. Perhaps she thought I would grow out of it or that I might be gay. If the latter, she never asked me about it (she would not have approved and surely saw homosexuality as immoral).

Although I did not yet understand the nature of my differences, I saw two movies during high school that had enormous emotional impact on me because of these differences. One of these was the 1958 science fiction movie *The Fly*. In a failed scientific experiment the protagonist becomes half man, half fly, ultimately committing suicide in a hydraulic press. Somehow I sensed that I shared an identity with this scientist and his ultimate fate. The other movie that powerfully emotionally affected me in high school was the 1932 horror movie *Freaks;* I identified with the deformed circus performers.

With the exception of Harry Potter (considered a freak by Muggles), I have never identified with any movie characters except for the fly and the freaks in these films. But it was many more years before I understood that the real monsters in the movie *Freaks* were not the physically deformed circus performers but the "normal" members of the circus who humiliated and degraded those performers.

It is difficult to express the degree of continued emotional pain, low self-esteem, and ultimately strong suicidal ideation that my gender discordance caused me while growing up and as a young adult. It was only at the age of forty, as I will come to later in this book, that I finally understood that I was transgender and was able to deal effectively with the problem. But growing up I was too confused to talk with anyone about it or to have any idea what to say. It just made no sense that I was a girl feeling like a boy. How could I ever utter something like that to anybody?

The cause of transsexuality is not known. Recent identical twin studies show a concordance rate of about 30 percent, so part of it may be genetic. Female animals and humans exposed to male hormones during fetal development exhibit masculinized behaviors. My mom was treated with a testosterone-like drug during the first trimester of her pregnancy, when she was carrying my twin sister and me in 1954. Back then synthetic steroid drugs had only recently been invented and doctors were experimenting with them to see if they could prevent miscarriage in certain high-risk pregnancies (my mother had had some slight vaginal bleeding early in her pregnancy, which was thought back then, wrongly as it turns out, to put her at high risk of miscarriage). Many women were treated with diethylstilbestrol but my mother was instead given a testosterone-like drug. I suspect that this drug may have masculinized my brain. If so, it did not affect my twin sister. (For most transgender people there is no history of abnormal fetal hormone exposure.) Fetal iatrogen exposure, however, almost always affects one twin but not the other.

Unfortunately I never developed particularly close feelings of affection for anyone in my family. Partly this is because I moved away from home at a young age. But I suspect it may also be because I was unable to bring myself to share the constant emotional pain I suffered from my childhood years onward. I blame myself entirely for this. I sorely wish that I had been able to open up to my mother about it. In recent years, it has been realized that as many as 1 percent of people may be transgender and there has been so much public education that parents are often able to open the discussion with their transgender children as early as grade school. Whenever I see videos of parents talking openly and supportively with their (pretransition) transgender children, I weep. Much progress is left to be made, but I marvel at how far the world has come in recent years.

MIT

I decided when I was thirteen years old that I wanted to go to MIT. My eighth grade science teacher had mentioned some research done there. I looked up MIT in the encyclopedia and it said that it was one of the best science universities in the world. No one in my family had ever gone to college; my parents had never heard of MIT; and my parents could not afford to pay anything toward my college tuition or expenses. But I was sure I was going to go there.

When it came time to apply to college my senior year (1971), I talked with my guidance counselor and told him of my plan to apply to MIT. Although I was the strongest science and math student in my class and had nearly perfect SAT scores, he assured me that I would not get in there and strongly encouraged me to apply to some local, less competitive schools. I later found out that the same guidance counselor had encouraged several of the boys in my class to apply to MIT even though I ranked higher academically than they did. This was long before my sensitivities about gender-based discrimination had been raised. In any case, I ignored his advice and I applied only to MIT early decision.

By early December in my senior year, I learned that MIT had accepted me and that, thanks to their scholarship and loan programs, I would be able to attend despite my family's low income. I sure was happy! In retrospect, I know that major American universities were only just starting to admit women in any appreciable number in the late 1960s and early 1970s. So although I was a more than qualified applicant, the barriers that talented women faced when applying to top colleges were lessening substantially around the time of my application in 1971. Although MIT admitted women from the start, their numbers were very low, perhaps a few per year, until the 1970s. When I arrived in the fall of 1972, I found that only about 5 percent of students at MIT were women. As I felt that I was a boy, however, I did not particularly notice this and it did not concern me.

I loved MIT. I found immediately that despite my differences, unlike in high school, I fit in. I was a science nerd just like everyone else. The quality of the courses that I took and the faculty that taught them were all beyond superb. MIT has long prided itself on having its best faculty teach the undergraduate courses and boy did I have stellar professors there. In my freshman year, as with all the students, I took physics, calculus, and chemistry. Although I worked hard, I did not find that I needed to pull all-nighters, and I enjoyed everything that I was learning. Though I got As in most of my classes, I was no longer close to the best student in every science class any more. But this did not matter to me at all. I was finally being challenged academically, able to take whatever course I was interested in—and I totally loved it. Despite my relative lack of sexual attraction, in college I had my first and only boyfriend, whom I lived with for several years. We were not a good match and eventually split; I preferred a single existence from that point on.

Because of my financial difficulties, I would sometimes run out of money and need to scrape by until the next monthly scholarship check arrived. Although I am sure there were others in my class with worse financial difficulties, these experiences have made me realize the challenges that many students from low-income families face in college. I was not surprised to see that although Stanford and some other universities now pay full tuition and stipend for low income kids whose families make less than $60,000 per year, few of these kids are accepted into Stanford each year. Those who do get in still face enormous financial challenges as they have many expenses beyond tuition and room and board to somehow pay for—and many of these kids must send funds home to help support their families. We need to do much more to help these kids!

MIT's course requirements for any given major at that time were not very extensive, which made it possible to try new areas of interest. I entered as a double major in chemistry and computer science, and came very close to fulfilling the major degree requirements in both of these areas. But although I did very well in these courses, my professors failed to notice my ability and failed to encourage my interest (I will say more about this later).

In my sophomore year, intrigued by artificial intelligence and the excitement in this research area at MIT, I took a course on the subject. The professor lectured about ongoing efforts in the field to make intelligent computer programs, for instance to decode visual scenes or understand language. This got me curious to know more about how the brain actually worked, so I signed up for a very popular course called introduction to psychology and brain science (the term *neurobiology* had not yet been coined).

The professor who taught this course, Hans-Lukas Teuber, was a gifted teacher. His course changed my life. As was true of many of MIT's professors, in Teuber's lectures he talked as much about what was yet unknown as about what was already known, weaving the very latest research into every lecture. He had spent much of his own career studying how injuries in different brain regions affected the functioning of the brain. From his lectures, I understood that studying patients could be a powerful approach to understanding the brain. By the end of the course I had decided to become both a neurologist and a neuroscientist. I don't think I quite understood yet how many more years of study and hard work would be needed to accomplish these goals.

I changed my major to neurobiology (I had to create that major, which did not yet exist—an option MIT provided) but soon decided to switch my major to life sciences when I realized how many premed courses I still needed to take. These courses included biology, biochemistry, genetics, and cell biology, and were again all outstanding. I vividly recollect for instance the wonderful lectures of Salvador Luria, who taught my biology course at MIT. Again, he skillfully interleaved basic knowledge with the very latest research advances. Both Teuber and Luria wrote strong letters of recommendation for me that helped me to win a desperately needed full scholarship from Bell Labs that financed my last year of college.

Going to MIT was an incredible privilege. It changed my life and opened many doors for me. But it was not by any means a perfect experience. Although MIT was finally admitting more women, we were in some important respects not receiving the same education as the men. In 1972–1976, the years I attended MIT, there were almost no women on the faculty, so women students did not see many role models. The course lecturers were almost all men and the research that was being presented was virtually all done by men. Sometimes there were overtly sexist remarks made by famous male faculty during their lectures. The Nobel laureate who taught my first physics course made overtly sexist remarks in lectures and showed nude pinups, causing me to transfer to another physics course. Also, although I was an outstanding chemistry and computer science student and did extremely well in many advanced courses on these topics, the faculty did not notice me or offer me research opportunities in their labs as frequently happened to the male students. Both chemistry and computer science have long been highly male-dominated fields and women have historically not been particularly welcome. Even in 2017, there are few women on Stanford's chemistry and computer science faculties.

In the artificial intelligence course I took at MIT, I was the only student to solve a very difficult question on the take-home final exam whose solution involved constructing a LISP program with nested subroutines that recursively called on each other. The professor announced in class that since no one had solved it, he was not counting it toward our grades. After class I went up to the professor to show him that I had solved the question. To my dismay, he sneered at me and said that my boyfriend must have solved it for me. I was offended because he was unfairly and wrongly accusing me of cheating. It was many years before I realized that his meaning was deeply sexist—he just couldn't believe a woman had solved the problem when so many men had been unable to.

I imagine that if I had been a male student my name might have been mentioned in class or that the professor might have encouraged my career in computer science, or perhaps offered me an opportunity in his or a colleague's lab. This is why I get deeply angry when famous men (like Larry Summers, whom I will come to below) espouse the idea that women as a group are innately less good at science than men but say that of course they do not discriminate against individual talented women. They fail to miss the basic point that in the face of pervasive negative stereotyping, talented women will not be recognized. Such negative stereotyping is not supported by any data and is deeply harmful to all women.

Indeed when it came time for me to find a lab to do undergraduate research, although I was an outstanding student, I struggled to find any MIT lab that would accept me. Equivalent male students did not have much difficulty finding outstanding labs in which to train. I finally found a young female professor who was willing to supervise me. But although I worked long hours in her lab for several years, I received little mentoring and it was a less than ideal experience. It is not surprising to me that with this kind of gender-based discrimination so many women in my generation were (and still are) dissuaded from careers in science. I suspect that these discouraging experiences overall had much less effect on me than most women because, as I have mentioned above, I did not see myself as a woman. In any case, I entered MIT full of passion for science and I left the same way. In the end, that's all that really matters.

MEDICAL TRAINING

I next attended Dartmouth Medical School from 1976 to 1979. Medical school generally takes four years, but Dartmouth managed to reduce it to three. As I wanted to do research training after medical training, the shorter time to an MD degree was appealing. Also, living in beautiful rural New Hampshire seemed like a dream to me. I enjoyed living there more than any other place that I have been.

All of the medical school classes were on the honors, pass, or fail grading system, and we got two afternoons a week off. Dartmouth was known for its strong basic science training and I found that all of the courses were superb. While in medical school, I took full advantage of the beautiful New Hampshire environment. In fall and spring I frequently went hiking or biking, and during the long cold winter I was generally out cross-country skiing on the many beautiful local ski trails. Once I entered my clinical rotations, time for these activities largely diminished, but I am very glad that I took some time during my first year and a half at Dartmouth to enjoy these things, as I look back on that time very fondly.

During medical training, I found that the barriers for women started to become even more glaring than they were at MIT. Dartmouth was one of the last colleges to go co-ed. Many colleges began admitting more women in the late 1960s, but Dartmouth only started to do this in 1972. So although my class was only 20 percent women, these included some of the very first women to graduate from Dartmouth College. I would often go to Dartmouth home ice hockey games and hear old alums on the benches behind me loudly complaining about how the school had gone to pot since they started to admit women. Perhaps not surprisingly, I once again found that there were still barriers for women in medical education.

In my first year anatomy class, the male professor liked to show slides with pictures of nude females. I was grateful when one of the male students went up to the professor after class to protest this. Disconcertingly, when women students asked a question after lecture, male professors often responded to a nearby male student. I was interested in doing research in a neuropathology lab. I found a professor who was willing to admit me to his lab. Unfortunately I soon found that the reason that this professor had agreed to take me into his lab was so that I would talk with his wife. He was not willing to actually teach me anything or involve me in his research. Similarly when I got to the clinics, I found that women students were largely ignored by many of the clinical faculty. I quickly realized that if I wanted to learn anything I had to be fairly assertive. It was hard to get past the feeling that women were largely not wanted or respected. Fortunately, although neurology was largely a male profession back then, I do not recall anything but a supportive neurology environment at Dartmouth.

One day during an endocrinology lecture, the professor taught us about a rare condition called testicular feminization, now called complete androgen insensitivity syndrome, in which XY individuals are born phenotypically female because they have a mutation in androgen receptors and thus are insensitive to testosterone. These patients lack internal reproductive organs. The professor said that these "women" were often not told of their diagnosis and XY karyotype. I was kind of in shock after the lecture as I thought that maybe I had this condition but was never told. It was many months before I was able to read my medical records to find out for sure that this was not the case and that my karyotype was XX.

As I became more knowledgeable about reproductive disorders in medical school, I spent time trying to research more what was known about gender identity and whether there was any relationship between my gender identity confusion and my reproductive organ anomaly. There was no evidence that fetal steroid hormone exposure ever caused Mullerian agenesis; moreover most patients with Mullerian agenesis had normal female gender identity.

Only recently had the term gender identity even been coined by John Money, a psychologist at Johns Hopkins. His view was that gender identity was completely socially constructed. To prove this, he was studying a patient who had lost his penis from a circumcision accident when he was an infant, and he had convinced the boy's parents to raise him as a girl. Money wrote many papers asserting that the boy was developing with a normal female gender identity (as it later turned out, none of this was true, and Money's theories were later completely discredited). So my confusion remained. If Money was correct, there was no reason why my gender identity should not be female despite my reproductive anomaly. In recent years, genetic mutations have been linked to Mullerian agenesis, so I suspect that such a genetic defect led to my abnormal fetal development, which led to my mom's first trimester vaginal spotting, which in turn led to her doctor giving her the testosterone-like drug, which in turn affected my gender identity.

At the end of medical school, I continued to be interested in both neuroscience and neurology. I had to decide whether to do research training next or to proceed with a neurology residency. I asked quite a few different professors. This was not very helpful as by the end of this inquiry, I realized that each of them had advised me to do exactly what they had done. I was not in a rush to complete my training, and I was still strongly considering being both a practicing neurologist and a neuroscientist. I decided to proceed with a medical internship (one more year) and neurology residency (three more years).

Because my mother had been diagnosed with metastatic breast cancer, I wanted to do an internship near to where she lived in West Orange, New Jersey. I selected as my first choice a Cornell-based program at North Shore University Hospital (NSUH) in Manhasset, New York, where interns and residents spent half of each year at NSUH and the other half at Memorial Sloan Kettering Cancer

Center (MSKCC) and New York Hospital (NYH). Both the medical and neurology training in this program were superb and it was only about a 45-minute drive to West Orange. Fortunately they selected me, and I began my internship in July 1979.

In those days, there was not yet any limit on the number of hours that house officers could work. As an intern I worked about 110 hours per week, sometimes more, and was on call every third night. On the call night, there was no sleep or, if I was lucky, at most one hour. Interns always used to debate if it was best to sleep for that hour or not (I took any sleep I could get!). On the other two nights, interns typically worked until midnight (and the days often began at 6 a.m.). Some interns were unable to handle the stress and soon left the program. As I had gone to a three-year MD program in a rural location, I found that I was not nearly as well prepared as my fellow interns who had trained in city-based programs. But I quickly caught up. It was the most intense year of my life. It is good that there are now strict restrictions on how many hours a house officer can work per week, limiting the time to eighty hours per week. After all, eighty hours is still barbaric!

After internship, I began neurology residency. Call was still every third night but with every progressive year, there was more sleep to be had as the most junior residents shouldered much of the night workload. Alas, during the second year of my residency, my mother's cancer was rapidly progressing. I transferred her to MSKCC where I could keep an eye on her and her care, but she soon passed away. She was only in her mid-forties. She was a really great mother, who raised four successful kids under extremely difficult personal and financial conditions. It is sad that she did not live to see us kids do well and get to know all her grandchildren.

Although neurological disorders can of course be devastating to the patients they affect, I loved every moment of my neurology training. My fellow residents were terrific as were most of my "attendings" at NSUH, MSKCC, and NYH. With only one exception, all of the neurology attendings were men. One of the chief neurologists was very hostile and disrespectful in public to the few women residents and I both experienced this and watched his treatment of others. My fellow (male) residents would often tell me of overtly sexist things he would say to them behind closed doors. But all of my fellow residents and nearly all the other attendings treated me with respect so, as always, I just sucked it up and did the best that I could do in my job.

I was lucky to have Jerome Posner as an attending physician while at MSKCC. He was as gifted a neurologist as I have ever known, a brilliant teacher, and a very kind man. I served as chief resident of neurology in my final year of training. Overall the training was superb. I did well and passed my neurology boards with ease. But as I completed my residency, I increasingly reflected that even as a fully trained neurologist, I could offer my patients few treatments to help them with their neurological injuries and diseases. This strongly drove my desire to move on to neuroscience research training.

DOCTORAL TRAINING

As I had just completed seven years of medical training, many people encouraged me to skip graduate school and go straight to a post-doctoral fellowship. This idea did not appeal to me, as I had not yet done any neuroscience research. Moreover, during the seven years of my medical training, the field of neuroscience was exploding. I was not ready to select a specific research topic on which to focus and wanted to have a bigger picture of neuroscience first. I therefore applied to PhD programs in neuroscience.

I applied to many graduate programs as I wasn't sure anyone would be interested in accepting a neurologist with no neuroscience research experience. To my surprise, nearly all of the programs that I applied to accepted me. I decided to go to the relatively new neuroscience PhD program at Harvard Medical School as it allowed its students to rotate in labs not only at the HMS quad (main campus) but in the nearby Harvard-affiliated hospitals, which tended to be more disease oriented.

To attend Harvard, I turned down a lucrative offer to join a neurology practice on the North Shore of Long Island. I had spent six months of my neurology training in an electrodiagnostics lab and was highly skilled at doing electromyography (EMG) and nerve conduction testing (NCT), a lucrative skill back in those days. But the lure of a high salary held little attraction to me. This particularly puzzled my dad who I recall saying "Let me see if I understand this correctly: you are going to turn down a $200,000 per year salary after seven years of expensive medical training to earn $6,000 per year as a graduate student and start all over again?" I gave him an enthusiastic yes!

I began my graduate training in 1983. Unfortunately the day that I began my graduate training was also the day that my college and medical school loans all came due. So for the first two years of graduate school, I was also moonlighting as a neurologist at New England Baptist Hospital every Friday night to Monday morning where I rounded on all the neurology patients, performed EMG/NCT procedures, performed neurology consults and the many needed procedures somehow not done during the week, and covered any neurological emergencies. It was exhausting. I found that I was not performing as well in my graduate lab rotations and courses as I was not getting enough sleep and never had enough time. My first two lab rotations did not go well. The summer of my first year quickly arrived and it was time to select a third rotation.

The patch clamp technique had recently been invented and I was very interested in learning it. One possibility was David Corey, a new faculty member who would be arriving in the fall. But his lab was not set up yet so I was considering another lab instead. I called David up (he was still at Yale) to discuss the possibilities with him. I told him the other patch clamp lab I was thinking about. He knew it to be a very weak choice, something I was unable to yet judge for myself. Rather than telling me not to go to that lab, he suggested that since I could work with him in the fall, why didn't I select another lab where I could learn other skills that might come in handy when I was learning to patch clamp, such as tissue culture. He suggested Linda Chun's lab as she was next door to where his new lab would be at Massachusetts General Hospital (MGH). His advice helped me to avoid a poor mentorship choice that would have greatly limited my career. This example illustrates the enormous role that luck plays in scientific training. Despite the importance of selecting good mentors, first-year graduate students are rarely ready to make this choice wisely.

I was fortunate that Linda Chun agreed to let me do my third rotation with her. She was also a very young faculty member who had only recently established her own lab, so she had lots of time every day to talk with me and guide me. I started to really love being in the lab. I had selected her lab not only because of David's recommendation but also because I was very curious about glial cells from my neurology training. Glial cells were still very mysterious back in the early 1980s, but it was clear from histology and pathology studies that they were the majority of cells in the human brain and that they changed their properties radically in most or all neurological injuries and diseases.

Although it was known that oligodendrocytes were the myelinating cells, the roles of astrocytes were particularly mysterious. It was not clear what the astrocytes normally did in healthy brains—they were assumed to be largely passive and to be support cells for the neurons—or in diseased brains where they changed their properties and became "reactive." But whether reactive astrocytes were helpful or harmful was not known. In Linda's lab, I learned how to dissociate brains into cell suspensions and then how to culture those cells, including the glial cells.

Linda had done beautiful work as a graduate student with Paul Patterson culturing and studying sympathetic neurons and their responses to nerve growth factor (NGF), and as a postdoc she had studied immunology. Now she was interested in studying whether and how glial cells interacted with immune cells. Methods to purify and culture central nervous system (CNS) cell types were still in their infancy in 1984 so I spent a lot of time during my rotation playing around with different ways of isolating and culturing neurons and glia.

Linda was a wonderful mentor. She was full of passion for science and she asked big, important questions. She used new methods as soon as they were available and was undaunted by high-risk work. I learned a great deal from working with her. I also enjoyed being at MGH because some of the most legendary neurologists to ever live, including Ray Adams and C. Miller Fisher, were still in active practice there. I had read many of their papers in my neurology training and it was thrilling to hear their eloquent discussion of patients at case presentations.

In the fall of 1984, at the start of my second PhD year, I began my fourth rotation in David Corey's lab. There was as yet no one else in his lab so, to my very good fortune, he also had much time to talk with me and teach me many things. I was still quite immature as a scientist and all those years of medical training had not helped my research abilities, so I really needed some serious mentoring. David turned out to be a phenomenally good and caring mentor. He was not only an extraordinarily talented and rigorous scientist, but he also cared very deeply about quality mentorship. I have tried to emulate this in my own lab, but I always feel that I fall far short of the standard he set.

Some of the things that he did were to go to neurobiology lectures with me and insist that I ask a question after the talk, and when I generated interesting data he would often tell me to go make an appointment with a particular professor he thought might be interested in it. He would always allow me to join him when he met with visiting scientists. Although I was initially very reticent about doing all these things, my shyness quickly diminished as I realized that all these professors were always kind and were enthusiastic to discuss interesting new findings; I always benefitted enormously. David's focus was always on mentorship and expanding me as a scientist, and he encouraged me to take Cold Spring Harbor laboratory courses and to go to many meetings. It also helped that he was a Howard Hughes Medical Institute (HHMI) investigator, as his lab was always well funded so I was able to try new techniques and whatever crazy ideas for experiments came to mind.

When it came time to teach me how to patch clamp, David was still setting up his lab, and the bullfrogs whose ears he studied to understand mechanical transduction of inner hair cells, had not yet arrived. He suggested we practice on glial cells, which I had learned how to culture in Linda Chun's lab. This promised to be deadly boring, as back then it was neurons that were thought to express all the interesting voltage dependent ion channels and neurotransmitter receptors whereas glial cells did not fire action potentials and thus were thought not to express ion channels or neurotransmitter receptors.

I quickly learned how to patch clamp, and to my surprise I soon found that, though they were not electrically excitable, both astrocytes and oligodendrocytes expressed a broad array of different ion channel types and even neurotransmitter receptors. I decided to join David's lab (Linda remained a wonderful co-advisor) and spent the next six years cataloging and describing the various types of ion channels in glial cells in vitro and after acute isolation as well as how their properties compared to their neuronal counterparts. It was technically challenging work, as patch clamping was still very new. It took much effort to figure out how to patch on to these cells in order to achieve gigaohm, tight electrical seals. In addition, the computer programs required to acquire and analyze patch data did not yet exist, so I was (most enjoyably) able to use my computer programming skills to write many of these programs.

When I joined the Corey lab early in my second year of graduate school, there were obstacles to getting much research done. I was still taking required graduate courses and still moonlighting so that I could pay off my loans. I was often sleep deprived and was frequently caught falling asleep during lectures. One Monday morning, after a particularly grueling weekend of moonlighting and little sleep, I dragged myself into work. David looked at me and, realizing I was exhausted, asked me why I was doing this to myself. I explained to him that I needed the money and he exclaimed, "Is that all?"

He proposed that I quit my moonlighting and instead be paid by him at a postdoctoral level since I already had an MD and he had many HHMI postdoctoral slots as yet unfilled. This would immediately raise my salary from $6,000 per year to $17,000, which was sufficient for me to make ends meet. So I quickly accepted his kind offer, which also enabled me to move from the HMS quad dormitory to a small, alas cockroach infested, apartment in Beacon Hill only a block away from David's lab. This change immediately resulted in more sleep, less stress, and my suddenly starting to do much better in my courses and in the lab. Although some physician-scientists are able to combine clinical and research careers, I realized that I was not a good multitasker and, although I missed practicing neurology, I never again tried to combine clinical practice with research.

Besides, I was enjoying the research too much. As happened to me at Bell Labs, I found that I did not want to ever leave the lab. I could be found there nights, weekends, and holidays. I slept as little as possible (my clinical training had taught me I could survive without much sleep) and never took vacations. Once, though, I was feeling particularly tired and decided to go to Miami Beach for a week holiday. I arrived in Miami around 4 p.m., fell into bed, and slept until the next morning. I then went to the beach for about fifteen minutes, decided I would rather be in the lab, and flew straight back to Boston!

As I tried to formulate my thesis proposal, I turned to the superb Harvard Medical School library to learn what was known about glial cells and what methods were available to study them. I read the work of the early neurohistologists and found that Cajal had wondered about glial cells and what they might do. He had concluded that until better techniques were available to study them, their roles would be a mystery. I read the more recent work but, as few techniques had materialized, I despaired at their incoherence and I wondered how I could make much progress. One day I stumbled across the recent work of Dr. Martin Raff, a professor at University College London who had also trained as a neurologist and then had done brilliant work as an immunologist. He had recently turned to using his immunology skills to start to dissect the types of CNS glial cells and his first papers about glia were only just being published in the early 1980s.

These papers were exceptionally elegant and beautifully written. Raff took advantage of the optic nerve as a simple part of the CNS and showed that optic nerve cultures contained not only oligodendrocytes (OLs) and two different types of astrocytes that he called type 1 astrocytes (1As) and type 2 astrocytes (2As), but also a new type of glial progenitor that he dubbed an O2As for oligodendrocytes-type 2 astrocyte progenitor cells (the O2As were later renamed OPCs for oligodendrocyte precursor cells when it became clear that there were few 2As in vivo, even though they are generated robustly under certain in vitro conditions). Most excitingly, he identified and generated a variety of antibodies that could be used to specifically identify each of these cell types: both

1As and 2As expressed the GFAP antigen, 1As expressed the Ran2 antigen, 2As expressed the A2B5 antigen as did O2As, and oligodendrocytes expressed galactocerebroside.

The culture conditions that his lab developed not only allowed for the study of all of these cell types but also enabled their development and maturation to be studied. Moreover all of the axons in the optic nerve came from retinal ganglion cells (RGCs) and as these axons could all be severed by a simple surgical procedure this was also a very powerful system to investigate the nature of neuron-glial interactions. I immediately realized that the optic nerve system would be the perfect system to investigate the nature of ion channel and neurotransmitter receptor expression by different types of glial cells as well as the nature of neuronal influence on glial electrophysiological properties.

By coincidence, at the same time that I was excitedly telling David about these papers, he happened to notice that Martin Raff was coming to MIT to give a lecture. When the time came for the lecture, David insisted on joining me, which surprised me, since David's research focus was on hair cells. He insisted that we go to the lecture early and when Martin Raff entered the MIT auditorium to set up his slides, we were the only other ones present. David told me to go up to him and introduce myself and ask him a question. I was still terribly shy as I was only at the start of my second graduate year. I said no, but David insisted and as he was sitting there watching me I had no choice. I shyly approached Martin, who was very kind and answered my question. After Martin's elegant lecture, David insisted that I go ask him another! I had lots of questions, so I am glad that David was there to make sure that I asked them. Soon after, David suggested that I write to Martin to ask him to be on my thesis committee. I could not imagine that Martin would agree to this, but to my surprise he kindly agreed.

When the time came for my qualifying exam, Martin was there and made many helpful suggestions. This began an active correspondence (via airmail from Boston to England as the internet had not yet been invented) in which I would write to Martin with questions or with my latest data. He would in turn answer my questions or tell me about recent advances in his lab. He was full of passion for science and by early in my graduate career I was already thinking that it would be wonderful to do a postdoctoral fellowship in his lab. But I doubted very much that he would ever accept me, as I would often hear him telling others that he did not have room for them to do a fellowship in his lab. But in the meantime he was like a third graduate advisor, in addition to David and Linda, as he generously provided so many helpful thoughts and suggestions for my research.

I will not detail all of my research findings as a graduate student but some of the interesting findings that I made were that 1As, 2As, OPCs, and OLs all expressed different complements of ion channel types, in vitro and in vivo (as judged by examination immediately after acute isolation), and that OPCs expressed functional AMPA glutamate receptors.[1] I found that the presence of serum, and even different lots of serum, had profound effects on which types of channels were expressed by glia in culture, which taught me to avoid use of serum in cultures for the rest of my life.[2] I found that even voltage dependent sodium channels were present in astrocytes, although these cells were not excitable, and by single channel analysis that the glial sodium channels had different properties compared to their neuronal counterparts, with different kinetics and voltage sensitivity.[3]

In collaboration with Linda Chun, I also developed a simple method to very highly purify retinal ganglion cells by immunopanning, although the purified cells quickly died in culture.[4] I was unable to figure out how to keep them alive to study their interactions with glial cells until many years later, as I shall describe later. I also showed that the primary ion channel type in oligodendrocytes was an inwardly rectifying potassium channel and that over development these channels rapidly localized to the ends of their processes.[5] These channels were also highly expressed by the 1As, although they were not present immediately after their generation but developed by about a week later unless the RGC axons were severed, in which case they never appeared suggesting that neurons were inducing their expression.[6]

Because I found that astrocytes and oligodendrocytes expressed high levels of an occult voltage dependent chloride channel, that became activated in excised patches,[7] I proposed that astrocytes accumulate potassium during neuronal activity not by spatial buffering but by accumulation of K^+, Cl^-, and water, as do muscle cells, but with the caveat that there must be a missing signal, probably neuronal, that activates these chloride channels and in turn the accumulation of K^+ for later return to the glia after activity was over. This possibility remains uninvestigated but I still think it is intriguing. The injured brain swells rapidly after injury and release of such a neuronal signal might lead to new insight about why this swelling occurs and how to better treat it.

Because I liked doing experiments, I was accumulating many papers worth of data, but I was stalling on writing it all up. I had never written a paper before and that seemed boring compared to generating the data. David grew increasingly frustrated. It didn't help that I tended to mess around with other experiments that had nothing to do with my thesis. One night in my third year I used a dounce homogenizer to isolate some nuclei that I labeled with a fluorescent nuclear dye so that I could visualize them and try to patch onto them to see if there were any ion channels in the nuclear membrane. I did the experiment late at night so David would not catch me, but all of a sudden, around 10 p.m., he suddenly appeared. He asked me what I was doing. I said, "Nothing, just the same old thing." He said, "Let me see." He sat down at the microscope, looked through the eyepieces, and . . . I was busted! He was silent for a minute. I could see he was not happy. He looked at me and said, "You know, someday, if you ever learn how to focus, you are going to be a great scientist." That was David: even in criticism he was supportive and kind.

By my fourth year, when no papers had yet appeared, David banned me from the lab. He sent me home with a computer and told me not to come back to the lab until I had written at least one paper. Not being allowed in the lab was a terrible agony, so I wrote that paper in record speed. By the time I graduated, I had written six primary research papers (five of which were published by *Neuron* and the first of which was the inaugural article in the first issue of *Glia*) and a long *Annual Review of Neuroscience* article on ion channels in glia,[8] as well as a few other reviews and commentaries. That seemed like a lot, but the patch clamp technique was new and there was a lot of low-hanging fruit.

Although I had learned a tremendous amount about ion channel biophysics, all of these papers shared the flaw that they were largely descriptive and did not really provide any really new insight into what the glial cells were actually doing in the brain. I always tell students that one good paper is more than plenty for a PhD or post-doc; just ask an important question, and take it one step forward. That was hard to do back in the 1980s, but with the tremendous advances in methodology, it is now something within reach of every trainee, and what makes running a lab now so much fun.

There were many difficulties in graduate school, technical stumbling blocks, which I eventually solved, in figuring out how to isolate glial cells viably, how to stably patch record from them, and so forth, which I will not belabor. At one point, in my third year, there was an HHMI site visit where some super famous scientists were rolled out to hear about what research was going on in David's lab. David presented to them both the hair cell stuff, which they rated very highly, and the glia stuff, which they were not so impressed by, to put it mildly. I decided perhaps I should work on hair cells after all. The lab was quite small and it was really amazing that David was even allowing me to work on glial cells rather than hair cells. So I decided to work on hair cells for a while, and even got my name on the author line of one hair cell paper.

I actually did not do any of the experiments in the paper but had suggested a simple way of isolating hair cell stereocilia for biochemical analysis that David dubbed "the bundle blot," which turned out to work rather well. But after a few months I found myself thinking again about glial cells. There were so many things I wanted to understand about their roles both normally and in disease. I decided that I would keep working on them anyway despite the very negative HHMI review. Although I didn't realize it back then, although the HHMI folks were quite correct that the work was largely descriptive, nonetheless I was developing methods and laying the groundwork that would allow me in the future to finally be able to answer questions about glial cells.

As always, David was highly supportive and encouraged my return to the glial work. Although I did not do much work on hair cells, it was an incredible privilege to be in the lab to watch David and his trainees elucidate so many of the fundamental mechanisms of hair cell transduction in technically versatile and brilliant experiments.

During graduate school, my gender confusion was increasingly bothersome. I had still not been able to talk to anyone about it and was increasingly feeling suicidal. I was beginning to contemplate specific ways that I might kill myself. This got sufficiently scary and I decided to go talk with a psychiatrist for the first time in my life. I had one appointment with a Harvard psychiatrist that lasted about ten minutes. I told him that I was feeling increasingly suicidal, but was too ashamed to admit the gender confusion to him, which he did not ask me about. All I recall him asking me was whether I was close to anyone (such as friends, siblings, or parents) to which I replied no. At the end of our brief chat he pronounced that "I was unable to love." I doubted this was the case, although now I am not so sure.

In any case, my desire to see him or any other psychiatrist again waned, although my suicidal ideation persisted. Eventually I got up the courage to mention my suicidal thoughts to David. I do not think he believed me because I did not appear depressed (and I do not think I was depressed, nor have I ever experienced a clinical depression), but he did make an appointment for me to talk with a social worker that he knew. I talked to her regularly for several months. Again I could not bring myself to talk about my gender confusion, but we did talk a lot about low self-esteem. Eventually I felt a bit better and I returned to dealing with my problems, as always, by burying myself in my work.

In graduate school, it seemed that gender-based discrimination was much less a concern than it had been during my medical training. Most of the professors were men, but there were at least some terrific women on the faculty. Back then many male professors had affairs with their graduate students and postdocs with impunity despite the harm to the trainees that often resulted; I was uncomfortably aware of many such relationships. It has been a great advance in recent years that most top universities, in the United States anyway, now have explicit policies that govern such relationships. There was only one episode that I feel was likely gender-based discrimination that directly affected me.

In my last year of graduate school, I applied for the elite career-transition award that would have funded my postdoctoral fellowship as well as providing funding for my own lab someday. Harvard was allowed to submit two candidates into the national competition, and I earned one of those two spots! When I met the senior Harvard dean to discuss the competition, he said to me, "I shouldn't tell you this but you are one of our two candidates. The other one [who was a man] has a far weaker application. You have six papers in top journals and he only has one, and your letters of recommendation are also much stronger. You are definitely going to win this award." But the other guy won. No doubt he was also very talented, but within a year or two he dropped out of academia to start a biotech company, so he must have quickly given up the award. Other than the question of fairness, it really did not matter to me as I found other sources of funding for my postdoctoral fellowship.

I feel fortunate that I was able to do my doctoral work in the neuroscience program at Harvard Medical School. It was a superb program in every way. By the end of my seventh year, David decided that it was time for me to graduate. I had successfully defended my thesis (I still recall his concluding line of his introduction, which makes me smile—he mentioned all my papers and then said, "This would be enough work to get Barbara tenure at most universities; let's see if it's enough to get her a PhD at Harvard"!). I still had many experiments I wanted to do so was not in any rush to leave. So David said to me that on such-and-such a day I would no longer be paid.

I took the hint and moved on to my postdoctoral work, but it was very painful to leave his lab. In many ways I saw David as a father figure. I admired him deeply as a scientist, mentor, and friend. He was very kind to me and taught me so much, spending endless hours rehearsing talks with me, editing my papers, and so on. I know that the unusually high quality mentorship that I received in his lab was the key to my success in science; he went on to train many other highly successful scientists (ten of them have Harvard faculty positions, eleven if you count a Harvard offer I was once made but turned down). When it was time for me to leave his lab, it ripped my heart out. On my last day I sat in the MGH courtyard and cried at having to leave. And at the goodbye party, for the first of two times in my life, I got drunk. I returned to my Beacon Hill apartment for one final night with the cockroaches and departed for London the next day.

POSTDOCTORAL YEARS

I did my postdoctoral work in Martin Raff's lab between the years of 1990 and 1993. One day during my fifth year of graduate school, he was visiting me at MGH and I worked up the nerve to finally ask him if I could work in his lab. As I feared, he immediately said no, he didn't think that was a good idea. I was crushed! Eventually he said that if I really wanted to do that I could visit his lab at University College London. He had a very tiny lab in a very old building with only a few postdocs and no graduate students. I couldn't believe that such a small lab had been the home to so many important research contributions in cell biology, immunology, and neuroscience. He finally agreed that I could have a postdoc position with him. I guess I failed to tell him that I needed a couple of more years to finish my PhD (in England they are typically only three to four years) so I felt very guilty (and still do) when I realized after finally arriving that he had held a bench for me in his tiny lab for a couple of years—but he never said a word to me about this.

I was concerned about how I was going to finance my postdoc, as I still had education loans to pay off and few fellowships financed training abroad. Fortunately I was awarded fellowships by both the National Multiple Sclerosis Society and the Life Sciences Research Foundation. Each of these fellowships offered an annual salary of $18,500 and allowed foreign training. When I looked up the NIH recommended salary scale based on years of training, it recommended a salary of $37,000. So I asked NMSS and LSRF if I might simultaneously accept both of their fellowships and to my great surprise they both readily agreed. This was a huge relief. The LSRF funding also provided a generous annual $15,000 amount to help finance the cost of experiments, which was also extremely helpful. The great generosity of these two organizations made my postdoctoral training possible.

During my first year in Martin's lab, I did not listen much to his advice. I worked hard and long hours on at least six different projects. One by one, each of these projects failed. When the sixth project failed at the end of my first year, I remember walking home that night to my nearby apartment in Kings Cross (it was in the middle of a "red light" district, which I had not realized when I rented it, so I was often propositioned for sex when I walked home after work at 3 or 4 a.m.). As I walked home that night, the soul-crushing thought occurred to me for the first time in my life that perhaps I was not cut out to be a scientist; perhaps I was not good enough. This was without doubt the low point of my career. But I didn't yet realize that I had actually learned a lot from all those experiments that didn't work. I started to listen more to Martin's advice.

Martin was just starting to become interested in why cells die. Up to that time it was thought that only certain types of specialized immune cells and neurons die, but Martin was starting to realize that this might be a much more universal property of cells, including glial cells. He proposed a heretical idea: maybe apoptosis, programmed cell death, is a universal property of all cell types and that to avoid apoptosis they needed to be constantly signaled by neighboring cell types not to commit suicide. I thought this was a brilliant idea. I realized that I might be able to adapt the immunopanning method to purify RGCs that I had developed with Linda Chun to purify specific types of glial cells. This would allow me to test their vulnerability to apoptosis as well as to investigate the nature of the cell types and signaling molecules that inhibited this death. A prior postdoc in his lab, Ian Hart, had noticed that OPCs and oligodendrocytes often underwent apoptosis in culture, and another of his postdocs, Sam David, had noticed that oligodendrocytes underwent apoptosis after a proximal optic nerve crush (when they were deprived of signals from RGC axons).

Given that Martin had already identified and generated antibodies that specifically recognized each of the optic nerve glial cell types, it was straightforward to develop an immunopanning method to generate pure OPCs, pure OLs, or pure astrocytes from the optic nerve.[9] Consistent with Martin's hypothesis, I found that each of these cell types, just like the purified RGCs, when placed into serum-free culture, quickly underwent apoptosis. But this apoptosis could be avoided if the purified cells were co-cultured with other cell types or specific peptide trophic factors released by these other cell types.[10] For instance, astrocytes strongly promoted the survival (and proliferation) of OPCs and the survival of newly formed oligodendrocytes, as did PDGF-AA, CNTF, LIF, IGF1, and forskolin (which increased their cyclic adenosine monophosphate [cAMP] levels). Moreover, when I looked at a normal developing optic nerve, I found that some newly formed oligodendrocytes were normally undergoing apoptosis (and within an hour their corpses were being phagocytosed by microglia).

Our experiments indicated that at least half of newly generated OLs were undergoing apoptosis during normal development.[11] When I cut the optic nerve, mass death of OL lineage cells ensued, but I could prevent much of this death by addition of exogenous peptide trophic factors such as PDGF and CNTF.[12] Altogether these experiments provided strong support for Martin's ideas and also suggested the model that newly formed oligodendrocytes competed for limiting amounts of trophic factors from axons and that if they did not find an axon to myelinate within several days after generation, as might happen in already fully myelinated territories, they would die.[13] We still don't know what the relevant axonal signals are, but a postdoctoral fellow in my lab, Lu Sun, has recently stumbled on a signaling pathway in developing oligodendrocytes that very strongly controls their survival and has potential to lead to these axonal signals.

I also found to my surprise that the rate of OPC proliferation in the developing optic nerve was strongly stimulated by the electrical activity of RGCs.[14] In playing around with the components of serum-free medium historically used to culture OL lineage cells, I stumbled upon the fact that, in the presence of PDGF to drive OPC proliferation, OLs failed to be generated when thyroid hormone was left out of the medium.[15] It had been known that thyroid hormone strongly stimulates CNS myelination and that hypothyroid children are hypomyelinated, but the mechanisms responsible had been mysterious. Our findings provided one mechanism by which thyroid hormone could stimulate myelination, and later postdocs in Martin's lab beautifully extended this work to show that it was highly physiologically relevant.

The work I did in Martin's lab led to seven first author research papers in *Cell*, *Nature*, and several other journals, as well as a number of review articles (today a paper in one of these journals usually contains about seven times more data than they did thirty years ago). It was a fun time, as the methods we developed allowed us to ask some fundamental questions about glial development. Martin and his lab had laid a powerful background that I was able to help build upon, and it also helped that the field was just beginning to uncover the fundamental peptide trophic factors involved so that many of these were available to us in purified or recombinant form for our experiments.

Every night when I went home, typically around 3 a.m. (though on Sunday nights I usually went home early so I could watch Dr. Who at 11 p.m.), I would leave data from my latest experiments on Martin's desk. And when he was in town, the next morning before I got started doing experiments, I would drop by Martin's office to discuss the latest data. I learned a great deal during these discussions about how to think about questions, results, experimental design, and so on. As with David, Martin was a model of rigor and integrity, and he always had a focus on the important unsolved questions, as well as creative and hypothesis-driven science. He believed that killing off the hypothesis was always the best way forward. Whereas at Harvard there was often a tendency to ask critical questions at seminars that verged on destructive, I learned from Martin that a much better way was to ask constructive questions in the form of suggestions, such as what would happen if you did a particular experiment.

It was amazing that, in addition to Martin's day job, he also had another job—serving as a primary co-author on the book *Molecular Biology of the Cell*. He was often away at meetings, and whenever he would return from one he almost always did what he called a "report back." The night before these he would review the detailed notes he made from every talk and synthesize them, and in his report back he would tell us (his postdocs) for each talk what important question it addressed, what had been known, and then what the new step forward was. These report backs would often last for several hours, but attending one of them was far more educational than actually attending the meeting. Of course the report backs were as much for Martin as for us, as they enabled him to retain and synthesize new knowledge that would ultimately end up in the next edition of his book. Whenever I came back from the annual Society for Neuroscience meeting in the United States, Martin would immediately ask me to tell him what was the most important thing I learned. This kept me focused on big-picture, question-oriented research.

Once again I was fortunate to have found such a highly generous and exceptional mentor. As in David's lab, the training I received from Martin has been pivotal to my successful career in science. It is a trend these days to suggest to graduate students that they accelerate their training by skipping their postdocs. I cannot help but feel that those who do this are usually making a big mistake—why be in such a rush that you lose out on the chance to expand your horizons as a scientist? An important thing about postdoc training is that it teaches you that you can start on a new problem and rapidly be doing useful experiments. In my experience, those who skip postdocs are generally more risk averse and decades later are often still working on almost the same question they focused on in their graduate work.

Martin and I did have one running argument for the three years that I was in his lab. I claimed that he was away at meetings three out of every four days, whereas he claimed that he was only away one out of four days. We used the same dataset—his desktop calendar—to arrive at our conclusions! In my last week in his lab, before he went home one night, I gave him one of my final papers for him to edit. Deep in the methods section, I had buried in the middle of a sentence the following words: "Dear Raff, it has been really wonderful being in your lab even if you have been away three out of every four days." The next morning, as always, I found the fully edited manuscript awaiting me on my desk. I immediately opened it to that page and found that he had crossed out the three and replaced it with a one.

On my last day in London, Martin held a goodbye dinner and, for the second and final time in my life, I got drunk. I had such a wonderful time being in his lab; as had been the case with David, I thought of Martin almost as a father, and it was heart-wrenching to leave. But I wasn't really leaving, as both David and Martin have remained good friends and advisors throughout my career.

STARTING OUT AT STANFORD

I was thirty-eight years old when it came time to look for a job. I received wonderful offers from Duke, University of Washington Seattle, University of California, San Francisco, and Stanford and I agonized over which I would accept, making three visits to each school. Having been raised and schooled on the East Coast, I liked the idea of living on the West Coast for a while and I narrowed down the possibilities to UCSF and Stanford. Each had wonderful faculty, students, and facilities.

But there was something about the neurobiology department at Stanford that appealed to me. It was a small faculty that felt almost like a family, and they placed a high value on quality teaching and mentoring. Clearly this department was a place where research was thriving. Each of the faculty members had identified an important question and had successfully advanced their work with depth. This appealed to me because I knew that understanding the roles of glia would be challenging and that I would need supportive colleagues.

I chose Stanford. I would be the only woman in the department when I started, as Carla Shatz had just moved her lab to University of California at Berkeley, but this did not concern me. Soon after I arrived, I was surprised to learn that one of the reasons I was given the job was that the dean had decided it was time to better diversify the faculty. Because of a financial downturn, he had closed off all searches unless qualified women candidates were found. The way I learned this was that my chair at the time was telling male applicants that they need not apply unless they had an orchiectomy—that is got castrated!

This chair was a little eccentric, so I did not hold that against the department. When I went to say hi to him on my first day at Stanford, he looked at me, scowled, and said only "the clock is ticking" (he was referring to the tenure clock, of course). That was my welcome to Stanford! In his own curmudgeonly way, I think he meant well and, although I was initially quite offended to learn that I had been hired in part because I was a woman, in retrospect I think that this dean did exactly the right thing. In those days the med school faculty was nearly entirely male and they were rarely inclined to hire women. Twenty-five years later, our medical school faculty is more diversified, but we still have a long way to go.

As I had hoped, the neurobiology department turned out to be a wonderful place to start a lab. My colleagues were all terrific, Stanford provided a generous startup package, and it was easy to obtain great graduate students and postdocs. Within a couple of years of starting my assistant professorship, however, I realized that although I had sufficient lab space to do electrophysiological studies and tissue culture, I needed more bench space to do molecular and biochemical studies. I did not have any benches at all. There were many unused lab rooms on our floor at the time, so I did not think this would be a serious problem. I asked my chair if I could have a bench or two. To my surprise he said no, and that was the end of the conversation.

As a talented young woman scientist, I was not infrequently getting inquiries from other universities about whether I might like to move. I always said no, but given the serious space problem, when the next inquiry came along, from the neurobiology department at Harvard Medical School, I decided to take a serious look.

The offer came through in my fourth year as an assistant professor. I was being offered a tenured associate professor position at Harvard, with an endowed chair, a lab three times bigger than my current lab, and a much higher salary. I don't think that my department ever seriously considered that an offer with tenure would come through as at the time Harvard had not tenured anyone in a great many years and there were only a few tenured women professors in the entire medical school.

I suspected of course that one of the reasons that Harvard was offering me this position was as an attempt to better diversify their faculty. But I just wanted to do whatever I could to advance my lab's research program. I left a copy of the offer on my chair's desk and told him not to bother matching the offer because I had decided to leave (I admit that by this point I was quite angry at my department's failure to give me the couple of benches I had asked for as it was seriously harming my lab's research). To my great surprise, colleagues in many departments as well as my own learned of the situation and came together quickly to help. Stanford quickly matched Harvard's offer, and I decided to stay at Stanford. I had given my word to Harvard that I would move if offered the job, but I had not realized it would take two years for a firm offer to materialize. By that time, an unexpected and serious health issue had emerged.

After being at Stanford for only two or three years, at the age of forty, I developed breast cancer. I foolishly never did breast self-exams and just happened to feel the hard, painless mass in my right breast one day when my hand brushed against it. Before I had the mastectomy, I asked my surgeon if he would also remove my left breast. I told him that I suspected there was familial breast cancer susceptibility since my mom had also been diagnosed with breast cancer at a similarly young age (this was a few years before BRCA1/2 gene testing was available). I also told him that I did not feel I should have breasts, mentioning this to someone for the first time in my life.

He was initially horrified at the idea of removing healthy tissue, but my oncologist agreed that I might be at genetic risk and so my surgeon soon agreed. A few years later, testing showed that I indeed had a BRCA2 mutation. By extraordinary luck, and even though I had had that mass for several years prior to the surgery, none of my lymph nodes or other tissues revealed any sign of metastatic spread. Later studies showed that prophylactic mastectomy greatly decreases the chance of new breast cancers in genetically susceptible patients. And I was greatly relieved to be no longer burdened with having breasts. Doctors, nurses, and friends all encouraged me to have breast reconstruction surgery, but there was no way I was going to let anyone put breasts back on me! I did not yet understand that I was transgender, but I felt enormous relief at having a body that more closely resembled my internal male identity.

There was one other unexpected obstacle that I faced as an assistant professor. Even though I won many junior investigator awards, I could not win an NIH R01 grant no matter how hard I tried. In the mid-1990s, NIH R01 grant applications were 25 single-spaced pages (not counting all the administrative parts) and very time consuming to prepare. It seemed there were several issues as to why I could not win a grant. The first was that there were rules for the proper construction of such a grant, but I had no idea what they were, and it did not occur to me to ask anyone. The second problem was that I was starting completely new projects in my own lab.

I wanted to move from glial electrophysiology and glial development to investigate actual glial function. But at this point I had no idea what these functions were, and I had insufficient preliminary data to support my proposals. I was eventually able to overcome these two problems as I got advice from colleagues and as my lab generated more and more preliminary data. But I was to learn that there was yet another problem that was much harder to surmount. My reviewers simply did not believe that glial cells served any roles in the brain beyond their traditional support roles. Even when my lab had published two papers in *Science* demonstrating profound effects of glia in inducing both synapse formation and function, my grant reviewers still insisted there was no way that this could be true.

Grant application after grant application continued to be rejected. And they were not just rejected, but they were spit on and stomped upon. In the early 1990s, NIH agencies including NINDS, the National Institute of Neurological Disorders and Stroke, were funding only proposals that scored in the top 7 percent or so. But my scores were usually higher than 50 percent (that is they were usually triaged and not even discussed), though after about seven attempts I had only gotten up to 43 percent.

Things started to get a bit stressful. My lab was running out of funding from junior investigator awards and setup funds. My "clock is ticking" colleague told me that my failure to get a grant was highly embarrassing and that I should not tell anyone about the low scores I was getting. I worried that perhaps my colleagues were thinking they had made a mistake after all in hiring a woman. I started to seriously consider dropping out of academia. With my clinical training, I could easily go back to neurology practice or do research in a biotech startup or pharmaceutical company. Indeed companies were not infrequently approaching me about the possibility. But I really wanted to understand glia so I held on. Soon, to my great relief, two unexpected miracles happened that solved the problem for me and put the fun back into doing science.

The first was that one day I got an unexpected phone call from Jonathan Pollock, a young program officer at NIDA (the NIH institute for drug addiction). Jonathan and NIDA were trying to understand the neurological mechanisms that lead to long-lasting craving for addictive drugs, even when drug use had been stopped. He happened to come across the *Science* paper mentioned earlier in which we had shown that neurons had little ability to form synapses unless astrocytes were present. This suggested to Jonathan the possibility that addictive drugs might act on astrocytes, rather than neurons, to enhance their synapse-inducing ability. This might create extra synapses in a circuit that might leave a long-lasting effect on circuit function.

He told me that he noticed in the acknowledgment section of the paper that I had not mentioned NIH funding. Was it possible that I did not have NIH funding for this "beautiful" work? I started to explain to him about my difficulty with grant funding and he quickly pulled up from his computer database that I had recently obtained a score of 43 percent on a grant submitted to NINDS. The NIDA council was going to meet in a week, he told me, and he could not promise anything, but if I would write him a paragraph about the relevance of our work to addiction then he would see what he could do. One week later, he called me to tell me that he had transferred the grant to NIDA and that it was fully funded.

That was twenty years ago, and ever since he has looked after not only my lab but also those of my trainees who continued this work in their own labs. It is fair to say that understanding of the active roles of glia at synapses has advanced in large part as a result of his support. Another program officer at the National Eye Institute (NEI), Michael Oberdorfer, similarly provided advice and help in obtaining funding for my lab's work on retinal ganglion cell axon regeneration. Without Jonathan and Michael's extraordinary support, I would certainly have dropped out of academic research as an assistant professor. Once my lab was established and we had generated much more preliminary data, it became easier to obtain NIH support. But those first five years were very tough.

I am glad that NIH has recently started new mechanisms to specifically help support research of new and early investigators. Without these mechanisms in place during these competitive times, I fear that we could lose many talented young scientists. The NIH is a great institution. It may not be perfect but the difficulty of its job is hard to overstate. There is never sufficient funding for all the deserving proposals, so its officials must navigate a very difficult path in making sure that the funds that do exist are fairly and appropriately distributed. They do a magnificent job.

In those first few years of my lab, when I was having so many grant rejections, I took it very personally. But now, after a long career, I can look back on it with more detachment and see that this may be the way it always is when young scientists start with new ideas, often seeing the same old things in different ways. It is not personal, but just the usual resistance that new ideas are met with, particularly when the investigator is young and not yet independently established. Young investigators need to hang in there and realize that things will eventually get better.

There was one more unexpected occurrance that greatly helped my lab. After I had had a lab for about five years, I received a phone call from Vincent Coates—a highly successful engineer who had made many brilliant inventions in nanotechnology. He founded and ran a successful Silicon Valley company called Nanometrics, based on his invention of the first scanning electron microscope that allowed silicon chip quality to be assessed. He and his wife, Stella, wished to philanthropically support neuroscience research relevant to neurodegenerative disease. He had read about some of my lab's work and he wanted to find out more. Vince visited my lab and I told him about the work we were doing on glial cells and how I thought it might be relevant to neurological disease. He and his wife made a very generous gift to my lab and have continued to support our work ever since. Alas, Vince passed away a few years ago from Alzheimer's disease, and I am sad that he did not live to see the way that the work he supported in my lab has indeed led to new drug targets for treating neurodegenerative disease.

In addition to the Coateses' support, over the last twenty-five years my lab has been the lucky and grateful recipient of funding from several other philanthropic foundations including the Fidelity Foundation, the Sheldon and Miriam Adelson Medical Research Foundation, the JPB Foundation, the Christopher and Dana Reeve Foundation, and the Cure Alzheimer's Disease Foundation. Without their generous support much of my lab's work, particularly the most high-risk work we have done, would not have been possible. It has been a great privilege to know these donors and observe their generosity first-hand. Their generosity has not only helped fund my lab's research and many others, but it has made possible the training of the next generation of young scientists who will continue this research. These wonderfully generous folks have greatly stimulated my own philanthropic interests. I have bequeathed my entire estate to my department at Stanford. I very much hope that it will help our faculty to sustain their cutting-edge research.

TRANSITIONING FROM BARBARA TO BEN

After about four years at Stanford, I was promoted to associate professor with tenure. One morning, I was reading a local newspaper, the *San Francisco Chronicle,* and came across and read with astonishment a four-page article about Jamison Green, a female to male transgender person and transgender rights activist. He was one of few openly transgender people at the time. In the article, Green described in detail his personal experiences with gender identity and to my surprise they mirrored my own very closely. This was the first time that I understood that there were others who had the same gender identity discordance that I had. It was also the first time that I had heard the word transgender.

The article mentioned the clinic of Don Laub, a Stanford plastic surgeon who was a Bay Area pioneer in helping transgender people. As I started to read more about other transgender people, I realized that I was likely transgender. I made an appointment to be evaluated at his clinic. It was the first time I was able to discuss my gender confusion with anyone. I met with Dr. Laub, as well as with an experienced psychologist who had worked with him for many years. The clinic concluded that I was transgender and offered to help me to transition from female to male.

At that time, transsexuality was still listed as a mental illness in the *Diagnostic and Statistical Manual of Mental Disorders*, a classification of mental disorders published by the American Psychiatric Association. Proponents of this view argued that it was wrong and harmful to help people change their sex. Did I have a mental illness? I did not think so. Moreover, reflecting on my experiences during psychiatry rotations during my neurology training days, my impression was that the incidence of serious mental illness was likely far higher in psychiatrists than in transgender people. So I did not see why they should get to categorize me as mentally ill! Moreover, I had been exposed to a testosterone-like drug during fetal development and my masculinization was consistent with relevant animal and human data.

I felt an irresistible desire to transition from female to male from the moment I was offered that possibility. But I thought about it for several weeks because I was worried about what the repercussions might be for my career. Even though I was already tenured and so did not have to worry about being fired—a frequent outcome for transgender people in other professions at the time (in many states, transgender people are still not legally protected from being fired)—there was much to consider.

I did not know of any successful transgender scientists, and I worried whether, if I transitioned, I would be able to get any more grants (it was already nearly impossible). Would new students or postdocs wish to join my lab? Would my colleagues reject me? Would I still be invited to meetings and so forth? Reading about the experiences of other folks in other professions who had transitioned, I strongly feared that a transition would end my career. For about a week, I was almost unable to sleep from the stress as I pondered whether I should transition or commit suicide. I finally decided to open up to three friends whose opinion I valued very much: David Corey, Martin Raff, and Louis Reichardt. For the first time, I opened up to them about my gender confusion and told them that I was considering changing sex. Did they think that the repercussions would be so bad that it would harm my career? To my great relief, all three were immediately and strongly supportive. Based on their support, I decided to transition. I sent out the following letter to my colleagues, family, and friends late in December of 1997 to let them know of my gender dysphoria and my decision to transition.

Dear friends,
I am writing to disclose a personal problem that I've been struggling with for some time. It is important for me to talk about it now in order that I can finally move forward.

Ever since I was a few years old, I have had profound feelings that I was born the wrong sex. As a child I played with boys toys and boys nearly exclusively. As a teenager, I could not wear dresses, shave, wear jewelry, makeup, or anything remotely feminine without extreme discomfort; I watched amazed as all of these things came easily to my sisters. Instead I wanted to wear male clothing, be in the boy scouts, do shop, play sports with the guys, do auto mechanics and so forth. Since childhood, I have been ridiculed and shunned by women and by men. At the age of 17, I learned that I had been born without a uterus or vagina (Mullerian agenesis), and that I had been exposed prenatally to masculinizing hormones. Despite plastic surgical correction of my birth defect, throughout my life I have continued to have intensely strong feelings of non-identity with women. Perhaps most disturbingly I feel that I have the wrong genitals and have had violent thoughts about them. My lack of female identity was brought home vividly to me recently after having bilateral mastectomies for breast cancer. This surgery, rather than being an assault on my female identity as it was for my mother, felt corrective as my breasts never seemed like they should be there anyway; the thought of reconstructive surgery has been repellant to me. Since the surgery, people who do not know me often call me sir, but that doesn't bother me either. It is not that I wish I were male, rather, I feel that I already am.

It would be difficult to describe the mental anguish that this gender confusion has caused me. Although I have never been clinically depressed, it has been the source of strong feelings of worthlessness, intense isolation, hopelessness and self-destructive feelings. I have never been able to talk to anyone about it because I felt so ashamed and embarrassed by it. It seemed that it must be my fault, that somehow I should be able to make myself be a woman. This is how things stood until two months ago, when I read in the newspaper about the existence of a gender clinic at Stanford. They found that I have a condition known as gender dysphoria. To my amazement, I learned that I am not alone and that my story is stereotypical of all of those who have this condition.

So what is gender dysphoria (also known as being transgendered or as gender identity disorder)? Those who have it feel from childhood a strong mismatch between their anatomical sex and their brain sex (gender identity). The cause is unknown but is thought to be biological, as some cases are clearly associated with a history of hormone exposure during development. Although it is not treatable by psychotherapy, the dysphoria is substantially lessened by a change in gender role. Treatment with testosterone induces normal male secondary sexual characteristics within 6 to 12 months. Most patients also opt for mastectomies, which I have already had, and hysterectomy, which nature has already done for me. In my case, testosterone treatment would have the added benefits of substantially lowering my chance of new or recurrent breast cancer, because it lowers estrogen levels, and would block the osteoporosis and menopausal symptoms that will otherwise follow when I have my ovaries removed because of my cancer susceptibility mutation.

After much reflection, I have made the decision to take testosterone. I will thus become a female to male transsexual. This has been a difficult decision because I risk losing everything of importance to me: my reputation, my career, my friends and even my family. Testosterone is a far from perfect solution; I'm still not going to be "normal" and social isolation will undoubtedly continue. But testosterone treatment offers the possibility that for the first time in my life I might feel comfortable with myself and not have to fake who I am anymore. I know that I am making the right decision because whenever I think about changing my gender role, I am flooded with feelings of relief. I will begin taking testosterone in February. A change in my appearance will not be visible for several months. By summer, I will begin to dress in men's clothes and will change my name to Ben. Throughout this process I will continue to work normally and to conduct myself in all ways as usual (except that I will only use single occupancy bathrooms). Although the idea of my changing sex will take some time for you to get used to, the reality is that I'm not going to change all that much. I'm still going to wear jeans and tee shirts and pretty much be the same person I always have been—it's just that I am going to be a lot happier.

Many transsexuals change jobs after their "sex change" in order to retain anonymity, but anonymity is obviously not an option for me—nor is it one I desire. I am tired of hiding who I am. More importantly I owe it to others who unknowingly endure this condition, as

I did, to be visible. Despite my 7 years of medical training, which I undertook to understand what was wrong with me, until 2 months ago I had never heard of gender dysphoria (oddly I somehow picked the right organ to study!). Had it not been for the transsexual who allowed himself to be the subject of the news piece I read, I would still not know about it. Sure I knew that sometimes there were male to female transsexuals but I had thought that these people were perverts. I am not a pervert; I don't seek pleasure—only relief from pain. Most transsexuals hide because of shame and fear, perpetuating ignorance and oppression about their condition. Their suicide rate is so high that some experts have called gender dysphoria a lethal disease. This is why I cannot hide.

In my heart I feel that I am a good scientist and teacher. I hope that despite my trans sexuality you will allow me to continue with the work that, as you all know, I love. I am happy to answer any questions.

Sincerely,
Barbara A. Barres

Despite support from David, Martin, and Louis, sending out this letter was still very scary. I found that my family was immediately supportive and so were all of my colleagues. I heard back from many of them very quickly. Here is the very first response that I received. It is from Chuck Stevens at Salk, a colleague I had long admired for his science and his wonderfully generous mentorship of so many young scientists:

Dear Barbara,
Thanks for the letter and the personal info. I have always been fond of the person in there and the gender makes zero difference to me—I expect you will find the same with all of your friends. Let me know when to change to "Ben."

Best regards, Chuck

All of the other responses I received were similarly supportive. And there it was: this shameful secret I had held inside of me for forty years was out, and within a few months I had transitioned to Ben simply by taking testosterone (mastectomies had already been done, but I did have my ovaries removed soon thereafter as they were a cancer risk because of my BRCA2 mutation; the testosterone prevented menopausal symptoms). My career went on as before without a hitch. I am not aware of a single adverse thing that has happened to me in the past twenty years as a result of my being transgender, but there was the immediate relief of all emotional pain as a result of my transition. Never did I think of suicide again and I felt much happier being myself (Ben), no longer having to pretend to be a woman. It is hard to explain how much relief I felt and how much happier I became. It was as if a huge weight had suddenly been lifted from my shoulders.

I should also say that Stanford as a whole was very supportive, including the provost, dean, and all my faculty colleagues. To be honest, I feared that some of the faculty in my department might be embarrassed by my transition. Back then the internet had only recently come into existence and there was still much ignorance about transsexuality. If they had any qualms they did not mention them and they were all completely supportive—even the curmudgeonly "clock is ticking" guy!

I would like to think that I eventually accomplished enough to fit in. I was elected to the National Academy of Sciences (NAS) in 2013. I was proud to be the first transgender scientist to be elected to the NAS and was upset when the academy president refused to mention this in the NAS press release on the grounds that the academy "had to deal with religious people." I was deeply disturbed by this as it denies LGBT people proper attribution for their accomplishments, particularly given the great need of LGBT students to be aware of successful role models. Fortunately other news writers soon mentioned it in pieces about me.

How did taking testosterone affect me? It is powerful stuff! There were some of the expected side effects such as increased sex drive for a while (almost like going through a second puberty) and the development of a male hair pattern. I was delighted to be able to grow a mustache and beard, but less thrilled with the rapid onset hair loss that began almost immediately upon start of testosterone (my photograph shows the extent of these effects). All cellulite quickly disappeared. Fat distribution changed from hips and buttocks to abdomen (but a lot stayed everywhere else too). I became much stronger even without doing any exercise. I had never been able to do a single pushup as Barbara, but after about six months of taking testosterone, I noticed that my triceps were beefing up. To my surprise, I was able to do ten pushups (and soon thirty, although I never really worked at it).

I did not particularly notice any change in mathematical, spatial, or verbal abilities, although I did notice on a test that was given to me before and after testosterone that my verbal abilities seemed a little worse and my spatial abilities seemed a bit improved. I still get lost every time I get in a car. Perhaps the most surprising and unexpected effect, though, was that I largely lost the ability to cry. Before testosterone I cried easily, and often cried myself to sleep because of the gender anguish. But after testosterone I found that I was almost entirely unable to cry any more. In response to some very strongly sad stimulus, perhaps I would shed a tear, but the feeling would almost instantly pass. Many other transgender men have told me this has happened to them also, whereas transgender women gain the ability to cry much more easily.

When I transitioned in 1997, it was thought that only one in about 20,000 people were transgender, but now, in 2017, it is thought that at least one in 200 people are transgender. LGBT people are often high achievers. Many LGBT people in my generation share growing up with a shameful secret and consequent low self-esteem. Perhaps this may drive us to work hard to succeed in order to prove our self-worth. Things are changing fast for transgender people. The internet has enabled relevant information to be easily researched and accessed, and the public is now being rapidly educated. TV shows often feature transgender characters, and transgender people can now serve openly in the military. There are still some battles being fought, such as gaining protection from being fired for being transgender, as well as bathroom protections, but the public is mostly sympathetic to and supportive of LGBT people, so I believe these battles will soon be won.

Most important, clinics are popping up to help trans children. As a result of public education, trans kids often self-identify, or are identified by their parents, even at grade school age. As they approach puberty, if their transgender identity persists, these kids can be treated with puberty blockers so they do not undergo permanent bodily changes inconsistent with their gender identity. Then when they are of age, at about sixteen years old, they can make the decision about whether they wish to transition. Up to now at least 40 percent of transgender people attempt suicide. I hope that kids who are able to transition early will be spared the anguish of growing up in the wrong gender with the wrong body, will be able to have more normal social and romantic interactions, and will not have to keep shameful secrets from their families. How I envy them!

I am happy to be an openly transgender scientist and to serve as a role model for young LGBT scientists. I hope that I have helped ease their way a little bit. LGBT students and postdocs at Stanford and other institutions frequently contact me to discuss whether or not to be open in their applications to various training programs. I always counsel them to be open about who they are, as it seems to me that currently the advantages far outweigh the risks. The vast majority of academics are highly supportive. It is very difficult to live life in a closet. It does not make sense to do this because of an occasional bigot. I have yet to have anyone tell me they regretted their decision to be open.

SCIENCE

DEVELOPMENT OF METHODS TO PURIFY AND CULTURE CNS NEURONS

In thinking about what projects to work on as I got my new lab started at Stanford, I felt that figuring out how to keep CNS neurons alive in culture was a high priority. If successful, this would provide an important tool for investigating neuron-glial interactions in order to better understand the functions of glial cells. Retinal ganglion cells made the most sense to start out with as I had already developed a method to highly purify them to greater than 99.5 percent purity by immunopanning from cell suspensions prepared from postnatal rat retinas. But oddly, unlike PNS neurons whose survival could easily be promoted in culture by specific peptide trophic factors (NGF for sympathetic neurons, BDNF for nodose ganglion neurons, CNTF for ciliary neurons, etc.), a wide variety of peptide trophic factors alone or all combined was frustratingly unable to promote RGC survival.

Anke Meyer-Franke joined the lab as one of my first postdoctoral fellows and decided to tackle this problem. Neurotrophic factors were just being identified and becoming available in recombinant or purified form. She tried testing a large variety of these peptide trophic factors individually and in combination on purified rat RGCs in serum-free culture, but always the purified RGCs quickly died of apoptosis over a one-day period. It had been reported that the survival of some types of PNS neurons could be promoted by intracellular cAMP elevation or by K^+ induced depolarization. Anke tried these also but once again found that they did not promote RGC survival.

However, when she combined several peptide trophic factors together—BDNF, CNTF (or LIF), and IGF1 (or high concentrations of insulin which activates IGF1 receptors)—together with either cAMP elevation (using forskolin or Chlorphenylthio-cAMP) or K^+ induced depolarization, now she found that about 65 percent of the RGCs survived for weeks or longer, extending beautiful dendrites and axons.[1] Anke found that this survival effect was mimicked exactly by culture of the RGCs on an astrocyte-feeding layer or just in astrocyte-conditioned serum-free medium. Oligodendrocyte-conditioned medium did not promote survival of the RGCs on its own, but Anke found when she combined it with her BDNF/CNTF/IGF1/forskolin cocktail, that RGC survival approached 80 percent. We still do not know what the identity of the trophic signal from oligodendrocytes is but we know that it is likely relevant in vivo because we observed that in mutant mice that lack oligodendrocytes, apoptotic RGCs are observed in the adult retina long after their period of normal cell death in the first week postnatal.

Anke wondered why the RGCs did not directly respond to BDNF when in side-by-side experiments where she purified nodose ganglion neurons, a type of BDNF responsive PNS (peripheral nervous system) neuron, BDNF was fully sufficient by itself to promote their survival. We worked with Louis Reichardt's lab at UCSF to investigate where the BDNF receptor, TrkB, was localized. Anke found that, although RGCs made TrkB mRNA and protein, it was not present on the RGC surface but rather in an intracellular store. When she cultured the RGCs in the presence of the cAMP elevator forskolin or K^+ elevation (which she showed worked by elevating intracellular cAMP within the RGCs), she was able to show using surface biotinylation experiments that TrkB was present on the RGC surface, inducing BDNF responsiveness.[2]

Subsequently, many other types of CNS neurons have been shown to respond similarly.[3] This was a remarkable and interesting discovery as it reveals a fundamental difference between PNS and CNS neuronal responsiveness to peptide trophic factors, with PNS neurons intrinsically programmed to be responsive but CNS neurons requiring extracellular signaling to be responsive. This may well be important in understanding the plasticity of neurons in the CNS. In fact Jeff Goldberg in my lab, while studying the mechanisms that control RGC axon regeneration, made the remarkable observation that the rate of axon growth in response to BDNF was enhanced almost ten-fold when the RGCs were stimulated at physiological rates of electrical activity by growing them on silicon chips.[4] He later showed in his own lab that this effect was mediated by elevation of intracellular cAMP levels, and he is currently investigating the relevant mechanisms.

Andy Huberman recently beautifully extended the significance of this work in his own lab by showing that RGCs regenerate their axons in vivo far more robustly and for a long distance, all the way to their targets, when they are electrically simulated either by visual stimulation or chemogenetic manipulation.[5] We still do not know, however, what the physiologically relevant signals are that promote RGC (or other types of CNS neurons) survival in vivo. RGC number is not appreciably affected in mutant mice that lack BDNF, CNTF or LIF, or IGF1. Later we will come to the more likely trophic signal.

When Martin Gartz Hanson was in the lab as a "post bac" for a couple of years, he wanted to highly purify and culture rat embryonic spinal motor neurons, another type of CNS neuron. He adapted a previous method of Chris Henderson's and found that while many different peptide trophic factors would individually promote partial survival, when he cultured them in cAMP elevators such as CPT-CAMP alone, he could promote the long-term survival of the majority of these spinal motor neurons[6] just as had been shown for purified PNS neurons by Gene Johnson and others many years before. Brad Zuchero in my lab later modified our immunopanning protocols in order to very highly purify DRG neurons for studies of myelination. All of our lab's methods to purify and culture neural cell types have recently been collected into a Cold Spring Harbor Laboratory manual[7] with step-by-step protocols.

I think that many neuroscientists believe that the survival signaling mechanisms that promote CNS neuronal survival is a question that has largely been solved since the pioneering work of Rita Levi-Montalcini and so many others showed that single peptide trophic factors are necessary and sufficient to promote PNS neuronal survival. We found that brain vascular cells secrete trophic activity that powerfully induces growth of CNS neuronal axons,[8] although it was not sufficient to promote their survival.

Gary Banker's group made a very important observation over forty years ago when they discovered that astrocytes secrete signals that powerfully promote the survival and growth of hippocampal neurons, and others have gone on to show that many types of CNS neurons can also be supported by astrocyte secretions. The ability of astrocytes to promote neuronal survival is likely critical in vivo—and not just in vivo—because in mutant mice in which neurons are generated but then astrocytes fail to be generated, CNS neurons quickly die. Identifying the astrocyte-secreted trophic factor(s) for neurons is a crucial and unsolved question in neurobiology. Although we have been able to take advantage of RGCs, which we can culture apart from glia in serum-free culture, and therefore study their interactions with glia (many of which I will come to shortly), it has long been frustrating that we cannot do similar experiments with hippocampal neurons or other types of CNS neurons.

We therefore decided to biochemically tackle the problem of the molecular identity of the neuron survival signals secreted by astrocytes. As they were soluble, we thought this should be straightforward. Madolyn Rogers began this work in my lab when she was a graduate student. It turned out to be very tough work, and soon Madolyn realized the reason: there were at least two different activities required to be present together in order for the hippocampal neurons to survive. One of these activities was a large molecule, over 100kD; the other molecule was much smaller—perhaps only about 500kD. When Madolyn finished in the lab, Jennifer Zamanian continued the project. To our surprise, Jennifer's experiments have revealed that the trophic activity that astrocytes secrete is entirely made of sugar chains, either chondroitin sulfate or heparin sulfate glycosaminoglycan chains.[9]

These chains are bound to a large variety of proteoglycans that are made and secreted by astrocytes. Unfortunately, addition of CS or HS to hippocampal neuron culture medium is not sufficient by itself to keep these neurons alive; they must be combined with the small molecule that we have not yet biochemically identified. As soon as we identify that molecule, it should finally be possible to have a completely defined serum-free medium with which most types of CNS neurons can be cultured. This will enable many interesting studies, and will provide a much better understanding of the nature of neuron-glial signaling mechanisms and their functional significance. It is remarkable that glycosaminoglycans, and not peptides or protein, are able to so powerfully promote CNS neuronal survival. This raises the question of the relevant CSPG/HSPG receptor, which will be another important question for future studies.

When she was in my lab, Ye Zhang took advantage of the methods that we have developed to culture highly pure neurons, astrocytes, and oligodendrocytes in serum-free medium to ask the question of what small (not protein) molecules they are secreting using a metabolomics approach. These were pilot experiments, but they were enough to show that each of these cell types is secreting a large variety of molecules; most have not yet been identified or characterized functionally. It is very likely that many interesting and important neuron-glial signaling processes are happening that we as yet have no clue of, nor do we know their functional significance in health and disease.

WHY DO CNS NEURONS FAIL TO REGENERATE THEIR AXONS AFTER INJURY?

The methods that we developed to purify and culture RGCs and optic nerve astrocytes and oligodendrocytes enabled us to branch out and ask some new questions. One fascinating question is why axotomized PNS neurons can survive and regenerate their axons, while axotomized CNS neurons die and fail to regenerate their axons. At the time I started my lab back in 1993, one claim was that retinal ganglion cells were being killed by glutamate excitotoxicity. Several other labs had found that when they injected glutamate or various glutamate receptor agonists into the retina, RGCs would quickly die. But we had found in our culture experiments that glutamate helped to depolarize RGCs and actually promoted their survival; we could never kill RGCs by dumping glutamate or glutamate agonists on them in culture.

Erik Ullian in my lab investigated what happened when he injected glutamate agonists into the retina. In contrast to many previous reports, he found that the RGCs were completely invulnerable. In fact many cells in the retinal ganglion cell layer were rapidly killed, but Erik showed that these were entirely displaced amacrine cells.[10] Remarkably, he found that amacrine cells and RGCs had the same amplitude glutamate currents, yet one cell type was clearly more vulnerable to excitotoxicity. What mechanisms make some neuron types more vulnerable to excitotoxicity than others remains an important unanswered question.

Another idea for why RGCs might die after axotomy, in analogy with what was known about PNS neuronal survival, was that the axotomy might block retrograde trophic signaling from target-derived trophic factors. Shiliang Shen, a research associate in my lab, decided to investigate this possibility. Other labs had found that simply injecting trophic peptides such as BDNF or CNTF was not sufficient to promote the survival of most axotomized RGCs. However, Shiliang reasoned that, as in culture, single peptide trophic factors might not be sufficient. Rather, several trophic factors, such as BDNF, CNTF, and IGF1, might need to be combined with a drug that would make RGCs traffic their trophic receptors to their cell surface, just as we needed to do in culture to keep RGCs alive. In fact, Shiliang showed that axotomized RGCs quickly lost their trophic responsiveness, but when BDNF and CNTF were injected into the retina together with CPT-cAMP, a cell-permeable cAMP analogue that is not digested by phosphodiesterases, the survival of axotomized RGCs was powerfully promoted.[11]

As mentioned above, however, it is far from clear that target-derived signaling is normally important for the survival of RGCs in vivo. In fact, RGCs are surrounded by astrocytes and astrocyte-like Muller glial cells, which likely make the critical trophic signals that keep the RGCs alive. Axotomy should not interrupt that flow of trophic signals within the retina. It is possible that RGCs die because after axotomy they become electrically silent (either because channels and receptors are down-regulated or because their intraretinal synaptic contacts are stripped off or degenerate) and their intracellular cAMP levels fall; they lose responsiveness to the astrocyte-secreted factors. As we shall come to later, the death of RGCs after axotomy turns out to be caused not by trophic deprivation-induced apoptosis but rather because the RGCs are murdered by nearby reactive astrocytes.

Nonetheless, having found ways of keeping RGCs alive after axotomy, either by expressing genes that block apoptosis such as bcl2 or by using the trophic approach reported by Shen and colleagues[12] or simply by studying their axon regeneration in adult rodents where it takes several weeks for all the RGCs to die, it is possible to explore the mechanisms preventing successful axon regeneration. Many previous studies have called attention to the role that inhibitory signals from reactive astrocytes and degenerating myelin play, which cause axonal growth cones to collapse and stop growing. Certain forms of CSPGs on astrocytes and Nogo in degenerating myelin are some of the inhibitory molecules that have been identified. Jeff Goldberg in my lab also identified semaphorin-5A as an oligodendrocyte inhibitor of axon regeneration,[13] and stunningly Alissa Winzeler and Wim Mandemakers in my lab identified the major myelin lipid sulfatide as the first lipid myelin-associated inhibitor of axon outgrowth.[14] They found that axotomized RGCs regenerate significantly better in mice whose oligodendrocytes lack this lipid. The relevant axonal receptor has not yet been identified and is an important question for future studies.

One of the reasons that PNS axons are better at regenerating is that degenerating myelin is rapidly cleared after axotomy, whereas that does not happen in the CNS, where degenerating myelin along with its inhibitory cues may persist for a decade or more after injury. Several young scientists in my lab have investigated the mechanisms of PNS myelin clearance with the hopes that this might lead to new insight into why myelin debris clearance occurs so slowly if at all in the injured CNS. Mauricio Vargas wondered if antibodies might play a role in clearing degenerating PNS myelin because the brain blood barrier might be prevented from helping clear CNS myelin. In a series of clever experiments, he took advantage of the JHD strain of mice that lack B cells so are unable to make any antibodies at all.[15] He found that these mice display a significant delay in macrophage influx, myelin clearance, and axon regeneration. He could restore rapid clearance of myelin debris by passive transfer of antibodies from naive WT mice or by an anti-PNS myelin antibody.

His findings show that degenerating PNS myelin is targeted by preexisting endogenous antibodies, and they demonstrate a role for immunoglobulin (Ig) in clearing damaged myelin during healing. They also suggest that the immune-privileged status of the CNS may contribute to failure of CNS myelin clearance and axon regeneration after injury. As we shall see later, other CNS myelin clearance mechanisms also fail after CNS injury. Mauricio's observations suggest a reason why Nogo antibodies delivered into the injured CNS by Martin Schwab's lab have consistently been found to promote axon regeneration better than occurs in mutant mice lacking Nogo. The Nogo antibodies may well be helping to promote myelin debris clearance by microglia or macrophages. It may be that any antibody that opsonizes degenerating myelin as Nogo does would be similarly able to help promote myelin debris clearance. To my knowledge this possibility has not yet been investigated.

Mauricio's experiments provided insight into how peripheral macrophages help to clear degenerating PNS myelin but, as his and other experiments had shown, the Schwann cells also phagocytose degenerating myelin but the mechanisms were not known. To better understand how Schwann cells clear myelin debris, Amanda Brosius-Lutz in my lab investigated this question. She developed methods to highly purify Schwann cells both before and after nerve crush, and then performed RNAseq experiments to identify candidate phagocytic pathways. She found that the major pathways that Schwann cells use to clear myelin debris are the Mertk and Axl pathways.[16] Purified Schwann cells in culture robustly engulf myelin debris, but Schwann cells that lacked both of these phagocytic receptors were unable to clear myelin debris at all. In vivo, the rate of myelin clearance was delayed when Schwann cells lacked these pathways, although ultimately myelin was cleared, presumably by macrophages and other immune system cells that serve a redundant role.

Amanda also showed that astrocytes in vitro robustly engulf myelin debris using these same phagocytic pathways, raising the question of why they fail to do so in the CNS in vivo after axotomy. As we will come to later, it turns out that reactive astrocytes induced by this type of injury lose their ability to phagocytose. As microglia have minimal ability to phagocytose myelin debris, and macrophages cannot enter distal to a nerve crush (they enter only at sites of blood-brain barrier breakdown), and as reactive astrocytes lose the ability to phagocytose myelin debris, it is now clear why the CNS fails to clear myelin debris. But these findings also suggest that if drugs could be developed that restored phagocytic ability to these reactive astrocytes, perhaps by reverting them back to normal astrocytes, that myelin debris clearance might well be induced after CNS injury.

As discussed earlier, much attention has been paid to the extrinsic mechanisms that inhibit axon regeneration, but when Jeff Goldberg was in my lab he wondered whether intrinsic mechanisms might come into play as well. Specifically, he wondered if CNS neurons as they aged, after development was over, might lose intrinsic capacity to regenerate. Other investigators had shown that embryonic neurons transplanted into an adult brain had substantial capacity to regenerate a long distance even through myelin pathways. So Jeff purified RGCs from different aged rats from embryonic day 17 (E17) all the way to adult (P60) rats and measured the rate of their axon growth in response to BDNF signaling in serum-free medium. He found that embryonic RGCs rapidly extended their axons until postnatal age 0, around the time when they normally reach their targets. But remarkably when he cultured E17 RGCs in serum-free culture they continued to rapidly extend axons for weeks, long past the time when they would have slowed down in vivo. But when he cultured P0 or older RGCs for prolonged periods they never regained the ability to grow rapidly again. He reasoned that some cell-cell interaction around P0 must trigger the RGCs to irreversibly lose their growth ability.

Surprisingly it was not target innervation that did this; rather, Jeff found that amacrine cells signal neonatal RGCs to undergo a profound and apparently irreversible loss of intrinsic axon growth ability. Concurrently, he found that retinal maturation triggers RGCs to greatly increase their dendritic growth ability.[17] His findings provide strong evidence that adult CNS neurons fail to regenerate not only because of CNS glial inhibition but also because of a loss of intrinsic axon growth ability. By performing gene profiling on RGCs of different embryonic and postnatal ages,[18] he identified some interesting gene candidates and is currently investigating these in his own lab. It is likely that epigenetic changes underlie this irreversible switch.

Of course, the critical question is even if adult RGCs can be induced to regenerate their axons, will they be able to find their appropriate target regions and form functional synapses to restore visual function? When Andy Huberman was in my lab, he took a variety of clever approaches to identify transgenic mouse lines, many made by the Gensat Project of Nat Heintz at Rockefeller University, which expressed green fluorescent protein (GFP) just in subsets of RGCs. There are at least twenty different subtypes of RGCs and Andy was able to identify mouse lines in which about five of these types were specifically labeled. These mice provided a powerful tool to investigate the specific target regions of each RGC subtype, to study how each of these subtypes developed their connections, and to identify the molecular mechanisms responsible.[19] Having identified the normal target regions of each of these RGC subtypes, in his own lab (and in other labs as well), Andy has taken wonderful advantage of these lines to demonstrate that they can indeed regenerate back to their appropriate target regions, make functional connections, and restore visual function.

For all these reasons, the possibility that CNS function may be repaired after injury is increasingly plausible, causing Jeff Goldberg,

Andy Huberman, and others to propose to cure blindness by eyeball transplantation. There is much work left to be done to make this a reality.

UNDERSTANDING OLIGODENDROCYTE DEVELOPMENT, NODE OF RANVIER FORMATION, AND MYELINATION

Having figured out how to purify and culture RGCs, we next wondered whether we could get them to myelinate with oligodendrocytes. Studies of CNS myelination have long been hampered by the lack of a rapid and robust myelinating culture system. We found that when we added oligodendrocytes to the RGC cultures that, although we did not get any myelination (more about that below), to our surprise, beautiful regularly spaced clusters of sodium channels appeared along their axons even when there was no contact between the RGCs and the oligodendrocytes because we used conditioned medium.

This was a surprise because in the PNS, it had been shown that direct Schwann cell contact was needed to induce nodal clustering of sodium channels, and at the time it was thought that similar astrocyte contact of nodes was needed in the CNS to induce sodium channel clustering. But there were no astrocytes present in our RGC-oligodendrocyte cultures. We found that oligodendrocytes were secreting a protein that induced this clustering, but we have not yet identified this protein. Oligodendrocytes were required in vivo for the formation of sodium channel clustering along axons because mutant rats that lacked oligodendrocytes developed almost no sodium channel clusters.[20] The requirement of oligodendrocytes for inducing sodium channel clustering revealed an unexpected new function for oligodendrocytes.

We also investigated the mechanisms by which oligodendrocyte precursor cells (OPCs) differentiate into myelinating oligodendrocytes. It is known that the timing of oligodendrocyte differentiation from OPCs in different pathways is precisely controlled, occurring generally right after a given axonal pathway has innervated its target area. Songli Wang in my lab found that OPCs highly expressed Notch1 receptors and that activating these receptors in culture strongly inhibited their differentiation.[21] Moreover he found in vivo that RGCs expressed jagged1, a Notch ligand, and targeted jagged1 protein to their axons, but he also found that this jagged1 protein was down-regulated from axons right around the time of optic nerve myelination. This suggested the hypothesis that target innervation induced RGCs to down-regulate jagged1 mRNA or protein delivery to the axon. This possibility has not yet been tested, but other labs showed soon after that premature and ectopic myelination in fact occurs in the CNS in the absence of Notch1.

What so precisely controls the timing of CNS myelination, and why it occurs only after target innervation, is a fascinating and very important unsolved question. The lack of target-derived signals may well explain why robust CNS myelinating culture systems have yet to be developed.

In order to better understand how oligodendrocytes differentiate from OPCs to newly formed oligodendrocytes to mature myelinating oligodendrocytes, Jason Dugas took advantage of gene profiling methods and examined mRNA isolated from pure cultures of OPCs at various time points along their differentiation over a one week time period. He found that they fully matured into oligodendrocytes normally despite the absence of neurons and other cell types in several sequential stages, each characterized by the appearance of distinct transcription factors.[22] These gene profiles led to the identification of Id2, p57Kip2, Dicer1, mir-219, and KLF9 as powerful controllers of oligodendrocyte differentiation.[23]

Ben Emery, working together with John Cahoy, did additional transcriptomic work to zoom in on transcription factors that might control the ability of oligodendrocytes to myelinate. His studies led him to identify a new gene (at the time called gene mode 98) now named Myrf, for myelin regulatory factor, as a transcriptional regulator required for CNS myelination.[24] In the CNS, he found that Myrf is specifically expressed by post-mitotic oligodendrocytes and encodes a nuclear protein containing an evolutionarily conserved DNA binding domain homologous to a yeast transcription factor. In mice lacking MRF within the oligodendrocyte lineage, premyelinating oligodendrocytes are generated, but they are unable to express myelin genes and fail to myelinate. These mice display severe neurological abnormalities and die because of seizures during the third postnatal week. Thus Myrf is a critical transcriptional regulator essential for oligodendrocyte maturation and CNS myelination. A fascinating aspect that Ben has discovered in his own lab is that Myrf is a membrane-associated transcription factor that autoproteolytically cleaves to directly activate myelin genes.[25] Perhaps this may imply as yet undiscovered mechanisms for regulating myelination.

In order to better understand the molecular basis of myelination by oligodendrocytes, we have tried to develop more robust and rapid in vitro myelinating cultures. Along the way we have found that astrocytes release signals, as yet unidentified, that enhance ensheathment of axons by oligodendrocytes.[26] But genuine myelin wrapping did not occur in these cultures until Trent Watkins found when he was in the lab that inhibitors of gamma secretase triggered rapid myelin wrapping as demonstrated by electron microscopy.[27] He tried gamma secretase inhibition as he thought this might inhibit Notch signaling that might be inhibiting the wrapping, but his studies instead showed that it was not Notch but inhibition of some other gamma secretase substrate in oligodendrocytes that is not yet identified. By time-lapse microscopy, he was able to directly observe the process of myelination in this culture system. To our surprise, he observed that when an oligodendrocyte decided to myelinate it myelinated all axons it was going to myelinate within several days in a critical window of time early after the oligodendrocyte differentiated and then never myelinated other axons again.

The same observation was made shortly thereafter by David Lyons by imaging oligodendrocytes myelinate in zebrafish in vivo. This suggests that an intrinsic genetic program is operating that drives oligodendrocyte lineage cells through their successive phases of maturation and that once an oligodendrocyte has myelinated it may never be able to myelinate again, which has important implications for understanding the failure of remyelination in the disease multiple sclerosis. Very likely generation of new oligodendrocytes from OPCs will be critical for successful remyelination.

But what is the molecular basis of CNS myelination? It has long been known that myelin basic protein is essential for myelin wrapping as shiverer mutant mice that lack myelin basic protein (MBP) generate oligodendrocytes that cannot myelinate. When Brad Zuchero joined the lab, he carefully examined the gene profiles of differentiating oligodendrocytes and noticed changes in actin control genes. He showed that the initial stage of process extension and axon ensheathment by oligodendrocytes requires dynamic actin filament assembly by the Arp2/3 complex. But surprisingly he found that subsequent myelin wrapping coincides with the upregulation of actin disassembly proteins, and that rapid disassembly of the oligodendrocyte actin cytoskeleton does not require Arp2/3. When he induced loss of actin filaments this drove oligodendrocyte membrane spreading and myelin wrapping in vivo, and he showed that the actin disassembly factor gelsolin is required for normal wrapping. Remarkably, he discovered that MBP is required for actin disassembly and that its loss phenocopies loss of actin disassembly proteins.[28] His findings provided new insight into the molecular mechanism of myelin wrapping and identified it as an actin-independent form of mammalian cell motility. Brad has his own lab now and is doing some very exciting work aimed at understanding how MBP induces actin disassembly and myelination.

DEVELOPMENT OF METHODS TO PURIFY AND CULTURE ASTROCYTES AND ELUCIDATION OF THE ASTROCYTE TRANSCRIPTOME

When I started at Stanford, I had developed methods to purify astrocytes from the developing optic nerve, but these did not work to purify astrocytes from the rodent brain. At the time, the way astrocytes were cultured from the brain was to make cell suspensions from a neonatal brain and culture them in fetal calf serum containing medium for a week or so, then shake off the top layer of microglia and oligodendrocyte lineage cells (and perhaps some neurons), and then culture the remaining cells in a mitotic inhibitor to stop growth of contaminating cells.

This method, developed by McCarthy and DeVellis (in my lab we call these astrocytes MD astrocytes), took several weeks, was not prospective, and worked only with newborn brains. It was not clear whether the resulting cells were astrocytes or some sort of glial progenitor cell, and clearly many neural stem cells were present as new neurons were generated in these cultures over time. These cultures were also highly contaminated by residual microglia. Moreover, the use of serum greatly altered the properties of the purified cells, making them reactive. We desperately wanted to develop a prospective isolation method that would allow us to isolate astrocytes from more mature rodent brain tissue.

We worked on this for many years without success. We just could not identify antibodies useful for immunopanning that specifically bound to the surface of astrocytes and not other cells. When John Cahoy joined the lab he decided to tackle this problem. He screened various lines of mice that expressed GFP off of an S100b promoter. He found one reported in the literature that looked promising. His immunostaining experiments showed that it expressed GFP only in astrocytes and in OPCs. That was significant because he was then able to develop a simple prospective purification method where he first immunopanned with antibodies to remove OPCs and oligodendrocytes and then used fluorescent-activated sorting (FACs) to purify the remaining GFP cells which he showed were highly pure astrocytes.

As a side product he also got highly purified populations of OPCs, newly formed oligodendrocytes, and more mature oligodendrocytes. After the sort, the remaining cells were highly enriched neurons. The new generation of Affymetrix gene chips was now available, and John, working together with Ben Emery, was able to generate spectacularly beautiful datasets. Their paper, "A Transcriptome Database for Astrocytes, Neurons, and Oligodendrocytes: A New Resource for Understanding Brain Development and

Function,"[29] is the most highly cited paper my lab has published. It is hard to understate what a useful roadmap this turned out to be not only for my lab but for many others. It provided a gold mine of information that has helped us to better understand glial function and neuron-glial interactions by allowing us to formulate new and testable hypotheses.

These datasets led us to finally be able to develop an immunopanning method to isolate rodent astrocytes from the brain. Being able to prospectively purify and culture astrocytes in serum-free medium was vital to better understanding their functions and interactions with other brain cell types. Lynette Foo tackled this project. She was able to analyze the datasets to identify highly expressed astrocyte genes that encoded for plasma membrane proteins not expressed by other brain cell types. With much hard work and ingenuity she figured out a simple way to immunopan and purify brain astrocytes. This was only half way there, however, because, as expected, the purified astrocytes rapidly underwent apoptosis in culture.

She found that vascular cells strongly promoted astrocyte survival in serum-free culture and that they did this in part by secreting HBEGF, which was an interesting candidate since it is an EGFR ligand, a receptor that astrocytes highly express. By adding HBEGF to the serum-free culture medium Lynette was able to keep the astrocytes alive. She also found that some developing astrocytes normally undergo apoptosis in vivo and that the vast majority of astrocytes contact blood vessels, suggesting the hypothesis that astrocytes are matched to blood vessels by competing for vascular-derived trophic factors such as HBEGF. Finally she showed that compared to the traditional MD astrocyte cultures, the gene profiles of her cultured immunopanned postnatal astrocytes (IP astrocytes) much more closely resembled those of in vivo astrocytes.[30] The simple purification and culture method that Lynette developed has been an invaluable tool for much of the astrocyte work our lab has subsequently done. Using this method, when Ye Zhang and Steven Sloan joined the lab, they were able to take advantage of RNA-sequencing technology to construct even better transcriptomes and a splicing database of glial cell types, neurons, and vascular cells in the cerebral cortex.[31]

ELUCIDATION OF ACTIVE ROLES OF ASTROCYTES IN SYNAPSE FORMATION AND FUNCTION

Having developed methods to purify and culture both RGCs and astrocytes, we were finally in a great position to ask the question: what do neurons do by themselves and what do they need astrocytes to do? When Frank Pfrieger joined the lab he asked whether astrocytes might play any role in controlling synapse function. In vivo the majority of synapses are ensheathed by astrocytes. The astrocytes have largely been thought to have relatively passive roles, isolating synapses from one another, helping control ion concentrations, and rapidly clearing released neurotransmitters. But might astrocytes also have more active roles at synapses?

To find out, Frank cultured RGCs in the absence and presence of astrocytes. In the absence of astrocytes, in serum-free medium with trophic factors for the RGCs, the RGCs looked beautiful extending dendrites and making axons, and they were electrically excitable if injected with depolarizing current. However, to our surprise, they displayed little spontaneous synaptic activity and had high failure rates in evoked synaptic transmission. But when co-cultured with astrocytes, either in direct contact or just in conditioned medium, Frank found that the frequency and amplitude of spontaneous postsynaptic currents were potentiated seventy-fold and five-fold, respectively, and fewer transmission failures occurred. Astrocytes increased the action potential-independent quantal release twelve-fold. Thus, RGCs in culture form inefficient synapses that require astrocyte signals to become fully functional.[32] This was quite a surprise to us because at the time it was thought that neurons autonomously expressed all the molecular machinery needed to form functional synapses.

What could account for this lack of synaptic activity in the absence of astrocytes? Were the RGCs failing to form synapses or were they forming synapses that were not functional? Erik Ullian took up this question next. By performing a variety of experiments including quantal analyses, FM1-43 imaging, immunostaining, and electron microscopy, he found that few RGC synapses form in the absence of astrocytes and that the few synapses that did form were functionally immature. But astrocytes increased the number of mature, functional synapses on the neurons seven-fold and were also required for synaptic maintenance in vitro. In addition, he found that in developing rodents in vivo most synapses are generated concurrently with the development of astrocytes. Thus Erik's findings demonstrated that astrocytes are actively involved in inducing and stabilizing CNS synapses.[33] Similarly Erik, together with Brent Harris, found that Schwann cells strongly promote synapse formation between purified spinal motor neurons in culture.[34]

These findings taken together strongly supported active roles for astrocytes in inducing synapse formation and strengthening synapse function. But in order to find out if they were functioning similarly in vivo, we next needed to identify the astrocyte-secreted molecules that were promoting synapse formation and function. Karen Christopherson took a biochemical approach to identify thrombospondins (TSPs)-1 and -2, which were expressed and secreted by developing but not mature astrocytes, as promoters of CNS synaptogenesis in vitro and in vivo. She found that TSPs induce ultrastructurally normal synapses that are presynaptically active but postsynaptically silent and work in concert with other, as yet unidentified, astrocyte-derived signals to produce functional synapses.[35]

Cagla Eroglu took up the question of what the identity of the neuronal thrombospondin receptor was. She identified it as alpha2delta-1, a voltage dependent calcium channel subunit of unknown function with a large extracellular domain, which was also the known receptor for the anti-epileptic and analgesic drug gabapentin. She showed that the VWF-A domain of alpha2delta-1 interacts with the epidermal growth factor-like repeats common to all thrombospondins, that alpha2delta-1 overexpression increases synaptogenesis in vitro and in vivo, and that it is required postsynaptically for thrombospondin- and astrocyte-induced synapse formation in vitro. She also discovered that gabapentin antagonizes thrombospondin binding to alpha2delta-1, powerfully inhibiting excitatory synapse formation in vitro and in vivo.[36]

Together these experiments demonstrated that astrocytes not only promote synapse formation in the developing CNS, but also were an important step forward in understanding the therapeutic mode of action of gabapentin. Cagla also identified another highly expressed astrocyte-secreted protein called Sparcl1 (hevin) as a strong promoter of excitatory synapse formation in vitro and in vivo[37] and found that Sparcl1 induces glutamatergic synapses by bridging neurexin-1alpha and neuroligin-1B.

Astrocyte-secreted thrombospondins and SparcL1 both induce the formation of structural synapses, but these synapses are postsynaptically silent. So how do astrocytes promote synapse function? Frank Pfrieger had shown in his own lab that astrocyte-secreted cholesterol powerfully enhanced presynaptic efficacy. When Nicky Allen joined the lab, she found that astrocytes also profoundly promoted postsynaptic function. Remarkably, she discovered that in pure cultures lacking astrocytes, RGCs express all of the AMPA receptor mRNAs (GluR1, 2, 3, and 4) and translate them all into protein, but this protein remains inside the RGCs and fails to get to the synaptic surface. Astrocytes secreted signals that rapidly induced all four AMPA glutamate receptors to get to the synaptic surface. She used biochemical fractionation of astrocyte-conditioned medium to identify glypican-4 (Gpc4) and glypican-6 (Gpc6) as astrocyte-secreted signals sufficient to induce functional synapses between purified RGCs. Application of Gpc4 to purified neurons increased the frequency and amplitude of glutamatergic synaptic events and Gpc4-deficient mice have defective synapse formation, with decreased amplitude of excitatory synaptic currents in the developing hippocampus and reduced recruitment of AMPARs to synapses. Her data identified glypicans as a family of novel astrocyte-derived molecules that are necessary and sufficient to promote glutamate receptor clustering and receptivity and to induce the formation of postsynaptically functioning CNS synapses.[38]

Remarkably, although astrocytes recruit all four AMPA receptors to the synaptic surface, GPC4 was able to recruit only GluR1, indicating that other astrocyte signals not yet identified recruit the other GluRs. Indeed, in her own lab, Nicky has already identified the astrocyte-secreted molecule that recruits GluR2. Together these findings reveal a remarkable complexity of astrocyte-neuron interactions that control synapse formation and function. There are many astrocyte-secreted molecules that control synapses awaiting discovery. Indeed while we have focused so far on excitatory synapses, another lab has shown that astrocytes also strongly promote the formation and function of inhibitory synapses.

ELUCIDATION OF ACTIVE ROLES OF ASTROCYTES AND MICROGLIA IN SYNAPSE PRUNING

The developing brain initially makes excess synapses and an activity-dependent process eliminates the weaker un-needed or inappropriate synapses. The mechanisms that eliminate synapses have until recently been largely mysterious. This was not a problem that my lab was initially working on, but our lab's first gene chip experiment led us to think about it.

At the time, we were trying to understand how astrocytes so strongly stimulated synapse formation. We wondered if maybe astrocytes simply induced expression of synaptic genes in RGCs. We tested this in a gene chip experiment by comparing mRNA levels in RGCs that had and had not been exposed to astrocyte-conditioned medium. To our surprise, there were few genes that significantly changed in levels, but all three mRNAs encoding the three subunits of the complement component C1q, the initiating component of the classical complement cascade, were highly up-regulated. This was a surprise as C1q and other complement components were not thought to be expressed by healthy brain tissue.

Developing brain tissue, however, had not been examined before. By immunostaining we found that C1q immunoreactivity was strongly localized to developing synapses throughout the CNS in the first week postnatal. As this is a time period when extensive synapse elimination is occurring and the known role of the classical complement cascade is to help mediate elimination of bacteria, apoptotic cells, and debris, we immediately hypothesized that the classical complement cascade was helping to tag unwanted synapses for uptake and elimination by microglial cells, which express high levels of phagocytic complement receptors.

Beth Stevens had just joined the lab and did some beautiful experiments assessing the role of this pathway in retinogeniculate synapse refinement. Indeed, she found that part of this refinement was impaired in C1q- and C3-deficient mice, strongly implicating the classical complement cascade.[39] Moreover when she imaged microglia during early postnatal development, she observed synaptic remnants, but these were greatly decreased in mice that lacked C3 or the microglial complement receptor.[40] Pruning of complement-coated synapses did not continue into adulthood.

In addition, Won-Suk Chung in the lab found that astrocytes also helped to mediate part of retinogeniculate synapse refinement by mediating synapse elimination.[41] This hypothesis had initially also been suggested by our transcriptome datasets, which showed to our great surprise that astrocytes were highly enriched in multiple phagocytic pathways including the MEGF10 and MERTK pathways. Up to that point, microglia had been thought to be the main phagocytic cells present in the brain. But Won-Suk found that astrocytes in the developing brain contained synaptic remnants of both excitatory and inhibitory synapses, but these were absent in mutant mice that lacked MEGF10 and MERTK (engulfment of synapses by astrocytes was independent of C1q). Phagocytosis of synapses by astrocytes continued into adulthood though at a slower rate than postnatally. Remarkably, Won-Suk found that electrical activity strongly controlled the phagocytosis of synapses by astrocytes. When he silenced electrical activity of RGCs in both eyes, little phagocytosis in the lateral geniculate occurred, but when only one eye was silenced the silenced synapses were preferentially phagocytosed.

These findings raise many questions. Why are some synapses engulfed and not others? And why are some engulfed by microglia and others by astrocytes? Are there other mechanisms of synapse elimination that do not depend on glia? How does activity control phagocytosis? Perhaps one of the most interesting questions is what is the role of the continued activity-dependent engulfment of synapses in the adult hippocampus and CNS in general? Is it possible that this engulfment is critical for the structural remodeling of synapses involved in learning and memory? Laura Clarke in the lab is investigating this question now. Given the abilities of astrocytes to control synapse formation, function, and elimination, it is clear that these cells are not passive support cells after all, but are critical components in the functioning and plasticity of neural circuits.

UNDERSTANDING HUMAN ASTROCYTES: IS THERE AN ASTROCYTIC BASIS TO HUMANITY?

As we found that rodent astrocytes were so strongly controlling synapse development and functioning, we increasingly wondered whether these properties would be shared by human astrocytes. The enhanced cognitive abilities of the human brain compared to other animals are generally attributed to evolution of neural circuits, but might the synaptic abilities of astrocytes have also evolved in beneficial ways that could enhance cognition?

Ye Zhang and Steve Sloan in the lab developed an immunopanning method to acutely purify astrocytes from fetal, juvenile, and adult human brains and were able to maintain these cells in serum-free cultures.[42] They found that human astrocytes have abilities similar to those of murine astrocytes in promoting neuronal survival, inducing functional synapse formation, and engulfing synaptosomes. In contrast to mouse astrocytes, though, they found that intracellular calcium in mature human astrocytes responds robustly to glutamate. They performed RNA sequencing to compare gene expression with rodent astrocytes and identified many highly expressed human-specific astrocyte genes whose functions remain unknown. They also found that, in comparing human astrocytes to mouse astrocytes, more changes in gene expression occur in astrocytes than the neurons.

Their work identified two specific stages of astrocyte differentiation: fetal brains contained only astrocyte precursor cells (APCs), which were highly proliferative cells that expressed only immature astrocytic properties, but between six months and twelve months after birth they found that the APCs differentiated into post-mitotic astrocytes with their fully mature pattern of gene expression. The timing of this astrocyte maturation overlaps exactly with the time window when synapse density in the developing human brain greatly increases.

Because human astrocytes share the abilities of rodent astrocytes to control synapse development and function, it is possible that some developmental disorders and diseases might be caused by defects in these astrocyte abilities. Steve Sloan in my lab therefore collaborated with Sergiu Pasca's group to investigate whether human astrocytes were generated in the induced pluripotent stem cell-derived 3D human cortical spheroid (hCS) culture system that Pasca had invented.[43] He acutely purified astrocyte-lineage cells from hCSs at varying stages up to twenty months in vitro and performed RNA sequencing to directly compare them to purified primary human brain cells. He found that hCS-derived glia closely resemble primary human fetal astrocytes and that, over time in vitro, they transition from a predominantly fetal to an increasingly mature astrocyte state. The hCS-derived astrocytes closely resemble primary human astrocytes and will be highly useful for future studies of development and disease. By comparing the properties of the APCs and mature astrocytes, Steve found that both astrocyte stages have equivalent ability to induce synapse formation but that the APCs engulf synapses at a vastly higher rate. This suggests that the emergence of a high density of synapses in the human cerebral cortex between six months and twelve months of postnatal age may not be the result of synapse formation induced by mature astrocytes but rather the result of a greatly decreased rate of synapse pruning.

What controls the timing of astrocyte maturation is an important unsolved question. If this maturation timing becomes abnormal, occurring either too early or too late, it might have irreversible consequences on the formation of neural circuitry.

DEVELOPMENT OF NEW TOOLS TO STUDY MICROGLIA

Dysfunctioning microglial cells are increasingly implicated in many neurological diseases, but little is yet know of their normal functions. A stumbling block has been the lack of tools to identify, purify, and genetically manipulate them apart from closely related macrophages.

Mariko Howe Bennett in my lab identified transmembrane protein 119 (Tmem119), a cell-surface protein of unknown function, as a highly expressed microglia-specific marker in both mice and humans.[44] She developed monoclonal antibodies to its intracellular and extracellular domains that enabled the specific immunostaining of microglia (and not macrophages) in histological sections in healthy and diseased brains, as well as isolation of pure non-activated microglia by FACS. This enabled her to construct RNAseq profiles of gene expression by highly pure mouse microglia during development and in adulthood as well as after an immune challenge. These profiles demonstrated that mouse microglia mature by the second postnatal week and suggested novel microglial functions for future investigation. Mariko is currently constructing an inducible TMEM119-Cre mouse line that, if successful, will provide an invaluable tool for manipulating microglia and studying their in vivo functioning in health and disease models.

Another great limitation in studies of microglia has been the lack of a serum-free culture system. Most studies of microglia have required the use of serum containing a medium to avoid microglial death, but the serum induces activation and alteration of the microglia properties. To better study microglia and the properties that distinguish them from other tissue macrophage populations, Chris Bohlen in my lab developed defined serum-free culture conditions to permit robust survival of highly ramified adult microglia. He found that astrocyte-secreted CSF-1, TGF-β2, and cholesterol and that together these three molecules strongly promoted microglial survival in serum-free culture.[45] However, with Chris Bennett in my lab, he also found that mature microglia rapidly lose their signature gene expression after isolation in culture but that this loss can be reversed by engrafting the cells back into an intact CNS environment. Their data thus indicate that the specialized gene expression profile of mature microglia requires continuous instructive signaling from the intact CNS. Identification of this microglia maturation inducing CNS signal is now an important goal.

Currently Chris Bennett in the lab has generated antibodies to human TMEM119 for isolation of human microglia so that their properties can be directly compared to mouse microglia. He is also generating a novel humanized mouse that will enable engrafting and study of human microglia so that their functions in disease can be better investigated.

STUDIES OF BLOOD-BRAIN BARRIER FORMATION

When Rich Daneman joined my lab, he decided to investigate the molecular mechanisms that regulate CNS angiogenesis and blood-brain barrier (BBB) formation, which were largely unknown at the time. He developed methods to highly purify and gene profile endothelial cells from different tissues, and by comparing the transcriptional profile of brain endothelial cells with those purified from the liver and lung, he generated a comprehensive resource of transcripts that are enriched in the BBB, forming endothelial cells of the brain.[46]

Through this comparison he identified novel tight junction proteins, transporters, metabolic enzymes, signaling components, and unknown transcripts whose expression is enriched in central nervous system endothelial cells and provided a valuable resource for further studies of the BBB. His experiments revealed an essential role for Wnt/beta-catenin signaling in driving CNS-specific angiogenesis and provided molecular evidence that angiogenesis and BBB formation are in part linked.[47] He found that the BBB is formed during embryogenesis as endothelial cells invade the CNS and pericytes are recruited to the nascent vessels over a week before astrocyte generation, and that pericytes are necessary for BBB formation.[48] He found that the pericytes induced the formation of tight junctions and vesicle trafficking in CNS endothelial cells, and also inhibited the expression of molecules that increase vascular permeability and CNS immune cell infiltration. Thus pericyte-endothelial cell interactions are critical to regulate the BBB during development, and disruption of these interactions may lead to BBB dysfunction and neuroinflammation during CNS injury and disease.

Overall his studies illustrated how each component of the BBB (tight junctions, vesicular transport, and transporters) is under separate control by different cell-cell interactions and molecular signaling pathways. By imaging the BBB, Dritan Agalliu found that these barriers were differentially affected during stroke, with stepwise impairment of transcellular barrier followed by paracellular barrier breakdown.[49]

UNDERSTANDING REACTIVE ASTROCYTES AND THEIR ROLES IN NEURODEGENERATIVE DISEASES

Reactive astrogliosis is characterized by a profound change in astrocyte phenotype in response to all CNS injuries and diseases. But it has been unclear whether reactive astrocytes are helpful or harmful. To better understand the reactive astrocyte state, Jennifer Zamanian in my lab decided to purify and gene profile reactive astrocytes from two different mouse injury models: ischemic stroke induced by middle cerebral artery occlusion and neuroinflammation induced by systemic injection of the immunostimulant lipopolysaccharide (LPS).

To our surprise, reactive astrocyte phenotype strongly depended on the type of inducing injury.[50] Reactive astrocytes in ischemia up-regulated many neurotrophic factors and thrombospondins, suggesting they might be helpful by promoting survival and repair, whereas reactive astrocytes induced by LPS up-regulated many classical complement cascade components, suggesting they might be harmful by driving synapse loss. We named these two types of reactive astrocytes A1 and A2 for their hypothesized "bad" and "good" functions after the M1/M2 macrophage nomenclature.

Work from Michael Sofroniew's lab had already strongly supported a repair-promoting function for A2 reactive astrocytes induced by ischemia, so Shane Liddelow in my lab decided to investigate the role of the LPS-induced A1 neuroinflammatory reactive astrocytes.[51] He found that microglia that were activated either by LPS exposure or by CNS injury induce A1 astrocytes by secreting Il-1α, TNF, and C1q, and that these cytokines together are necessary and sufficient to induce A1 astrocytes.

He was able to create cultures of pure A1 reactive astrocytes by simply adding these three cytokines to their serum-free medium, which allowed him to directly compare the function of normal astrocytes with A1 reactive astrocytes in vitro. He found that A1 astrocytes lost most of the normal astrocyte functions, losing the ability to promote neuronal survival, outgrowth, synaptogenesis, and phagocytosis. But he also found that A1 reactive astrocytes gained a new function: they secreted a neurotoxin that rapidly induced the death of neurons (as well as axons and synapses) and oligodendrocytes.

Because A1s are rapidly induced in the CNS after acute injury, these findings suggested the possibility that axotomized CNS neurons die after axotomy not because they are deprived of retrograde neurotrophic signals but because they are murdered by A1s. Shane tested this by investigating whether death of axotomized RGCs could be prevented by preventing the formation of A1 astrocytes after injury by inhibitory antibodies to Il-1α, TNF, and C1q or in mutant mice deficient for all three cytokines. Remarkably he found that RGC death was entirely prevented when A1 formation was blocked. These experiments provided strong evidence that A1 reactive astrocytes are responsible for the death of axotomized RGCs. As Shane found that A1 astrocytes are abundant in degenerating regions of most major human neurodegenerative diseases including Alzheimer's, Huntington's, Parkinson's, amyotrophic lateral sclerosis, and multiple sclerosis, his findings suggest that A1 astrocytes may actively drive neurodegeneration in these disorders.

By developing drugs that prevent A1 formation, that revert A1s back to normal resting astrocytes, that convert A1 into A2 reactive astrocytes, or that block the A1 neurotoxin, in the future it may be possible to block or greatly decrease neurodegeneration. At present, Kevin Guttenplan in the lab, working with Shane, has used biochemistry to highly purify and identify a candidate neurotoxin.

A working model consistent with these observations is that neuronal dysfunction or injury leads to release of a signal that activates microglia, which in turn induces formation of A1 reactive astrocytes, which in turn secrete a toxin that kills specifically only the injured neurons (or oligodendrocytes). Our findings indicate that the A1 neurotoxin does not kill healthy neurons. By only killing injured neurons, this mechanism avoids innocent bystander killing of unharmed neurons. By removing only injured neurons, their synaptic inputs would be freed up to wire onto other nearby neurons, which might help to preserve circuit function. It is unclear why a mechanism would have evolved to kill neurons and oligodendrocytes, but analysis of the gene profiles of A1s indicates that they are activating strong antiviral and antibacterial defense programs. Thus A1s may have initially evolved to fight infection, but in the aging brain, or after brain injury, induction of A1s may be harmful.

Why is the aging brain so vulnerable to neurodegenerative disease? Alexander (Ali) Stephan, when he was in my lab, stumbled upon what may be an important clue. We had found that the classical complement cascade was targeting developing synapses, but specific antibodies to C1q for immunostaining and for biochemical purposes were not available. So Ali made a monoclonal antibody to mouse C1q. He found that it worked well for immunostaining (and unlike other available antibodies did not stain C1q-deficient mouse brain).

To our surprise, he discovered that as the mouse brain aged, its synapses became highly immunoreactive to C1q, beginning first in the hippocampus at only a few months of age; then gradually this synaptic staining spread throughout the CNS.[52] But unlike what occurred at developing CNS synapses during the first postnatal week, he found that the classical complement cascade did not become activated in the normal aging brain, and there was no evidence that synapse loss was occurring. Although C1q-coated aging synapses are not engulfed by microglia, Ali found that they were functionally impaired. The extent of this synaptic C1q accumulation with normal mouse and human brain aging was remarkable as Western blots indicated that C1q protein was increasing with aging about one-hundred-fold.

Ali's findings raise the question of why synaptic C1q increases with aging. As C1q is a lectin-like protein that generally binds to debris, dying cells, and foreign substances, an interesting hypothesis is that C1q is binding to "senescent" synapses that are exponentially building up during normal brain aging. Synapses might become senescent because they are turning over less rapidly, and indeed we found that the rate of engulfment of synapses by astrocytes decreases with brain aging.[53] Moreover, Won-Suk Chung found that the rate of synaptic phagocytosis by astrocytes was fivefold slower in vitro and in vivo in a human ApoE4 background compared to ApoE2 (ApoE3 was in between).[54] In addition he found that transgenic mice expressing human ApoE4 accumulated C1q more quickly than did ApoE2 mice. His work lends support to the idea that the buildup of synaptic C1q with age may indeed be the result of slower synapse turnover by astrocytes and more rapid accumulation of senescent synapses. Perhaps development of drugs that stimulate the ability of aging astrocytes to healthily turn over synapses might someday ward off the normal cognitive decline of aging.

Together these findings raise the possibility that the vulnerability of the aging brain to Alzheimer's disease and other neurodegenerative disorders might be in part because of the exponential rise in synaptic C1q levels, which places them at great risk for any "second hit" that might activate the classical complement cascade and lead to unwanted synapse loss. That is, could a normal developmental mechanism of synapse elimination become aberrantly activated in the adult brain, triggering progressive synaptic neurodegeneration?

This idea seemed likely, as soon we had discovered that the classical cascade was targeting developing synapses, because many studies had documented strong complement activation in a large variety of human neurodegenerative diseases, including Alzheimer's disease, where synapse degeneration is a prominent feature. Moreover our RNAseq studies showed that neurons lacked the high levels of multiple complement inhibitor proteins that most other cells in the body express. We began by investigating the DBA2J mouse model of glaucoma in collaboration with Simon Johns's lab.[55]

Although RGCs do not die in this model until almost a year of age, we were startled to find that C1q was highly up-regulated and localized to RGC synapses in the inner plexiform layer of the retina by two to three months of age, accompanied by marked synaptic neurodegeneration by four months of age, long preceding RGC death. Mutant mice that lacked C1q were very strongly protected from RGC death and optic nerve degeneration.[56] Similarly, in collaboration with several other labs, we found early complement-mediated degeneration of synapses as a very early sign of pathology in many other mouse models of neurodegenerative disease including Alzheimer's disease,[57] frontotemporal dementia, and spinal muscular atrophy.[58] As with glaucoma, in all of these models, C1q inhibition or deficiency was strongly neuroprotective.

What second hit might activate the classical complement cascade and trigger synaptic neurodegeneration? The most likely possibilities are triggers of A1 reactive astrocytes, because our RNAseq studies demonstrate that genes encoding classical complement cascade proteins including C1r, C1s, C4, C2, and C3 are among the most highly up-regulated A1 genes and would fuel the classical complement cascade (C1q mRNA and protein are highly present in microglial cells even in normal brain tissue). In addition oligomeric beta amyloid is a strong activator of the classical cascade. Triggers of A1 formation include acute and chronic neural injuries as well as immunostimulants such as lipopolysaccharide that may normally be produced by acute and chronic bacterial infections. A low level of A1 reactive astrocyte formation even happens with normal brain aging.[59]

As we found that A1s were the predominant type of reactive astrocytes in most or all major neurodegenerative diseases including Alzheimer's disease, there is likely ample complement present to drive the classical complement cascade and thus microglia-mediated synapse loss in these diseases. The neurotoxin released by A1s, which is not itself a complement component, likely additionally helps to drive neurodegeneration. Thus the classical complement cascade and A1 reactive astrocytes (and their toxin) are important new therapeutic targets that should be tested for efficacy in neurodegenerative disorders. Although septic encephalopathy and hepatic encephalopathy do not involve neurodegeneration, the possibility that A1s contribute to the encephalopathy is an important one for future investigations.

FOUNDING A BIOTECH COMPANY

Having realized that the classical complement cascade was an exciting new therapeutic target for treatment of Alzheimer's and other neurological diseases, around 2006 I started to talk with major pharmaceutical companies about the idea of making a drug to inhibit the cascade. Over a several year period, I found that I was meeting a wall of resistance.

Some people thought it would be too hard to make this drug; others thought it might have side effects. It was true that some C1q-deficient mice and humans had lupus, but in most cases this was mild, and in any case there was (and is) no evidence that acute C1q deficiency in adults would cause lupus. In any case, lupus is a very common drug side effect that is well managed by drug holiday when needed. C1q and C1s seemed to me the best targets for inhibition, as only the classical arm of the cascade would be blocked, leaving the other two arms and their immune functions intact.

After a few years, I realized that if a drug were going to be made, I would have to become more actively involved. I started talking to my friend Arnon Rosenthal about starting a new biotech company to make complement inhibitors. Arnon, long a highly successful neuroscientist at Genentech and then the successful founder and CSO of Rinat Neuroscience Corporation, had just sold Rinat to Pfizer. I had gotten to know Arnon while serving on the Scientific Advisory Board of Rinat and was impressed by what he had accomplished. In only six years, while leading Rinat, he had generated monoclonal antibody therapeutics that inhibited nerve growth factor for treatment of neuropathic pain, beta amyloid for treatment of Alzheimer's disease, CGRP for treatment of migraine, and PCSK9 to reduce LDL cholesterol.

We decided to cofound Annexon Biosciences in 2011 and obtained $500,000 in seed funding from Fidelity Bioventures. Arnon again led these efforts and was successful in making a monoclonal antibody that strongly inhibited both mouse and human C1q and thus was highly useful for efficacy studies in mouse models of disease. Later we were fortunate to work with Ted Yednock as Annexon's Chief Scientific Officer. Ted, previously CSO of Elan Pharma and inventor of the blockbuster drug Tysabri for multiple sclerosis, had great expertise in both neuroscience and immunology. Under his leadership, and that of Doug Love, CEO, we were successful in obtaining larger series A and B venture capital investments of $34 million and $44 million.

Doing this amount of fund-raising took a substantial amount of my time for several years, but it is very nice to see the effort finally moving forward smoothly. Our C1q antibody has so far shown impressive efficacy in multiple mouse models of neurological disease, including neuromyelitis optica, Guillain-Barre syndrome, spinal muscular atrophy, and Alzheimer's disease.[60]

As of now, having so far tested our antibody safely in mice and primates, we are currently testing its safety in human volunteers and hope to start our first clinical trial in 2017. We will start by testing its efficacy in Guillain-Barre syndrome and glaucoma, but our ultimate goal is to be able to test its efficacy in major neurodegenerative diseases including Alzheimer's disease, frontotemporal dementia, and Huntington's disease.

ADVOCACY

MENTORING YOUNG SCIENTISTS

It has been a very great privilege to mentor so many young scientists (see Trainees in the Barres Lab). I did not realize when I started my own lab at Stanford that this was going to be, by far, the most rewarding part of the job. This is not to say that the process of scientific discovery has not been continuously thrilling, because it has been. But it is even more exhilarating to watch young people develop into independent scientists and to play some role in guiding that process. Indeed, the process of scientific discovery and mentoring young scientists is completely interwoven.

In an academic lab at a top university like Stanford, the principal investigator (PI) is not the one doing the experiments and not even the one having most of the ideas (if I am lucky I get to make an occasional suggestion that is not instantly thrown under the bus). I can honestly say that the vast majority of my graduate students and postdoctoral fellows have been far more talented than I ever was. I have written before about my approach to mentoring,[1] so will not repeat those thoughts here. I have always felt that I was incredibly fortunate in my training to have had such exemplary mentors for my PhD and postdoctoral work.

I have tried to emulate their practices when it comes to mentoring, but often feel that I am not coming close. I find, like my mentors, that my natural tendency is to be as hands off as I possibly can and to allow my trainees to be as independent as possible. I tend to make suggestions for possible starting points when new trainees join the lab. Sometimes my suggestions are taken, sometimes not. In any case, these starting points soon evolve into something different and very often the trainee thinks of something better. As long as they work in the area of neuron-glial interactions, it fits well into the general lab environment.

I did have one postdoc, Andrew Huberman, who never worked on glial cells at all and was completely independent from his start in my lab. I consider Andy my "paying it forward" to help relieve my continued guilt at not having worked on hair cells as a graduate student. It is very nice that he is now an associate professor in the lab next door to mine!

Overall, as I have previously written, I feel that mentoring young scientists is something that involves great generosity.[2] It is a great challenge to stay at the leading edge of science and maintain grant funding. But to do this and still be highly generous to your trainees is even more challenging. That's why those scientists who manage to both be great scientists and great mentors are real heroes to me. Some wonderful examples include my own mentors, David Corey and Martin Raff, as well as Steven Kuffler, Seymour Benzer, Corey Goodman, Lily and Yuh Nung Jan, Marc Tessier-Lavigne, Bob Horvitz, David Baltimore, Louis Reichardt, Sol Snyder, Mike Greenberg, Bill Newsome, Richard Axel, Cori Bargmann, and Chuck Stevens (and of course many more).

TRAINING YOUNG SCIENTISTS ABOUT HUMAN BIOLOGY AND DISEASE

Being interested in disease, I had an unusually prolonged training period prior to starting my own lab. After college, I did three years of medical school (usually it's four years but Dartmouth once featured a three-year program), four years of internship and neurology residency, seven years of graduate school, and three years of postdoctoral training. Seventeen years of very lowly paid and long hours of hard work! I would do it all again because I loved every moment of it.

My training about neurological disease drove much of my life's research on glial cells and their roles in disease. But seventeen years of training is far too much to ask of young scientists who want to study disease. But what is a better way? I think as a community we need to put more thought into this. MSTP (MD-PhD) training, paid for by the NIH, has long been the main way that physician-scientists are generated. I enjoy having MSTP students do their graduate work in my lab because I find that I can engage with them in meaningful disease-oriented research in a way that typical neuroscience PhD students cannot generally do.

But MD-PhD training suffers from a serious flaw. Most of these trainees, by the time they finish their MD-PhD training and their residencies (and clinical fellowships) find, as I did, that it is too difficult to be simultaneously a successful physician and a scientist. To be sure, some manage to do it, but most do not. Nationally only about 30 percent of MD-PhDs continue to do research. Moreover, for those who do continue to do research, they have often done an accelerated PhD training period and generally skipped postdoctoral training (or at most have done a short research fellowship). It is very difficult therefore for them to compete for NIH funding and to do research at the highest level.

As a result, just as was true twenty to thirty years ago, we are applying the results of basic science discovery far too slowly to develop new treatments for disease. Many diseases have never been worked on by outstanding scientists. I like to use hepatic encephalopathy as an example. When the liver fails, the brain fails. Why? No one really knows, and only a few, frankly not so good, labs have ever worked on the problem.

There are countless other examples. When I began my lab at Stanford twenty-five years ago, working on disease was seen as a second-class activity. This is no longer the case. At least half of graduate students express strong interest in studying disease and, given recent scientific technological advancements, it is now possible to do first-class research on human biology and disease. Unfortunately we are not teaching most graduate students about human disease, or at best only superficially in a course or two.

As an attempt to begin to rectify this problem, thirteen years ago I introduced a master's of science in medicine (MOM) program at Stanford University to teach entering PhD students intensively about human biology and disease. What I did essentially was re-create the Markey Foundation program that had run at Harvard Medical School for PhD students for about six years during the 1980s when I was a graduate student there.

As with the Markey program, PhD students who take the MOM program essentially take the first one and a half years of basic biomedical science courses with the MD students, while delaying most of their PhD course work for one year. These courses include anatomy, histology, physiology, and pathology. In short, the MOM students add on what amounts to about a year of extra training time in return for an intensive knowledge of human biology and disease. We have taken about five MOM students per year (determined by funds availability; several times more graduate students apply into MOM each year). I have watched MOM training transform the education of every PhD student who has taken it. It has not only intensified their interest in disease but enabled them to seriously study disease. In nearly every case it has altered the choice of PhD thesis lab chosen.

But this is not enough. We are not reaching enough students and most schools do not have a MOM program. Furthermore, the MOM courses are not designed for PhD students. We need to change this. I would like to see a "Khan Academy"–like series of high quality MOM courses, designed for young scientists, freely available to all who wish to learn about human biology and disease. Moreover such a website might contain an area where physicians can tell young scientists what the most important unanswered questions and medical needs are from their point of view. Discussions could be posted about these problems and needs that might stimulate new research and advances. Perhaps a medical foundation like the Howard Hughes Medical Institute or the Chan Zuckerberg Initiative will step up and help.

I finished my training in neurology almost thirty-five years ago. Since then, there are still no substantially new treatments for stroke (other than clot busters), neurodegenerative diseases, or glioblastoma. For multiple sclerosis, there are some new immunosuppressive drugs that are highly risky and there is still a need for drugs that promote remyelination. Progress is glacial.

When one looks at efforts that have gone into treating some common neurological diseases, such as stroke and Alzheimer's disease, things are surprisingly disheartening. There have been over one thousand failed clinical trials to treat stroke, mostly all versions of "save the neuron from excitotoxic death," and similarly there has been one leading hypothesis for Alzheimer's disease treatment—lowering beta amyloid—and many largely failed trials focused on that.

Something is broken. Normally in science when one disproves a hypothesis, one moves on to another hypothesis. Why is this not happening in our understanding and treating of neurological diseases? There is a desperate need for new (and better) scientists to engage in these problems.

HELPING WOMEN IN SCIENCE

When I was young, I did not believe that any barriers would hinder my career as a woman in science. The story told at MIT and elsewhere was that academia was a meritocracy. It never occurred to me to doubt this. MIT even had a bulletin board where they talked about how they admitted women from their very beginning. What it didn't mention is that it accepted only an occasional few until the 1970s. Even when it was implied that I had cheated on that computer science exam because "my boyfriend must have solved it for me," it was many years before it even occurred to me that this was sexism.

I see this same belief in meritocracy in young women today; the idea that their womanhood confers barriers generally occurs only as they reach mid-career and see less competent men being promoted or given leadership positions while they are passed over. When I look at mid-level and senior women at Stanford, I do not think that most are thriving the same way that their male counterparts are. They are all too aware of the barriers and they are resentful (sadly, one woman put it this way in a Stanford survey: "I feel like if I failed to stop showing up for work, no one would notice").

The few exceptions are highly successful women who, also with few exceptions, tend to deny the existence of barriers for women.

Many young women would like to believe that these battles were successfully fought long ago. But much evidence says this is not so. For instance, recently there was a news article saying that three out of four senior women scientists at Salk Institute were suing because they feel they have been systematically denied the same space and financial resources that their male colleagues have enjoyed, reminiscent of the same battle that Nancy Hopkins waged at MIT almost twenty years ago.

In general, my perception is that most of my male academic colleagues are well-meaning and strongly believe that it is truly a meritocratic system for both men and for women. They are unaware from their own experience of the many barriers that women continuously face. The best explanation I have found for why it is so hard for men to understand that gender-based barriers truly exist comes from Shankar Vedantam's book *The Hidden Brain* (2009). Vedantam talks about an experience he had swimming in the ocean, not realizing that the tide was with him. As he swam, he felt stronger and more confident. But when he tried to swim back to shore, he found that the tide was against him, his confidence left him, and it was very difficult to return.

It was only changing sex at the age of forty and experiencing life from the vantage of a man that I finally came to be fully aware of these barriers. I have written about these experiences in my essay called "Does Gender Matter?"[3] and so will not repeat them here. But as my transgender colleague Joan Roughgarden so wisely summarized: "Until proven otherwise, women are presumed to be incompetent whereas men are presumed to be competent."

All transgender people, whether male or female, share a common anger at the very different way that society treats people simply based on their gender. A counselor who works with people who have recently transitioned once told me that her most difficult challenge is helping male to female transsexuals understand that their suddenly vastly lowered social status is not because they are now transgender but because they are now women.

My experience of being differently treated as a woman and then as a man, even though I was the same person, is the reason I was deeply angry when Larry Summers, then president of Harvard, proclaimed that one of the reasons that few women were getting tenure in science and engineering at Harvard under his leadership was that women were innately less able than men.

My essay "Does Gender Matter?," a detailed response as to why I disagree—and more importantly all the scientific evidence that compellingly speaks against his view—was published by *Nature* in 2006. I made the following points in it: There is no compelling evidence for *relevant* innate gender differences in cognition. There is overwhelming evidence for severe gender prejudice. Both men and women often deny gender-based bias; we all have a strong desire to believe that the world is fair. When faculty tell their students that they are innately inferior based on race or gender they are crossing a line that should not be crossed—the line that divides responsible free speech from verbal violence. In a culture where women's abilities are not respected, women cannot effectively learn, advance, lead, or participate in society in a fulfilling way.

I was stunned by the response to my commentary. It was covered by most major newspapers around the world; hundreds of TV shows and radio stations asked to interview me; I received eight book offers and hundreds of invitations to speak on this subject (and I continue to receive them). I gave one talk on this subject, titled "Reflections on the Dearth of Women in Science," at Harvard University on the condition that it be posted on the web for anyone to watch (the video can be found on YouTube and the PowerPoint slides, and notes with references under each slide, can be found at http://www.memdir.org/video/ben-barres-dearth-of-women-in-science.html). *Please* watch this talk or read through the slides.

In response to my *Nature* commentary,[4] I received over three thousand emails, the majority from women telling me of terrible experiences of gender-based discrimination (and in some cases serious sexual harassment by their male professors) that had hindered their careers. I provide here two of these messages that particularly strongly struck me. The first strikingly illustrates the relative neglect that women experience throughout their lives, a neglect that most men may not realize women typically experience:

Dear Ben,
Just wanted to say thanks so much for coming forward with your experiences. . . . I was a Harvard student. I remember strongly a meeting I had with the poet Adrienne Rich. I came away from the meeting feeling shocked—I realized it was the first time I'd felt truly taken seriously as a person.

Julie

Another of these messages was from Dr. Nalini Ambady, who alas passed away in 2013 from leukemia. I do not think she would mind my sharing her message here:

Dear Dr. Barres,
I very much enjoyed reading your thoughtful article in *Nature* and have attached a couple of papers from my lab that indicate quite clearly that sociocultural stereotypes affect the performance of both adults and children (as young as 5 years of age!). Interestingly, mine was

one of the first tenure cases that came before Summers at Harvard in 2002. He ruled against the Psychology department's positive recommendation. This was, of course, well before he made his infamous remarks and before it was revealed that he had disproportionately ruled against women in tenure decisions in his first years in office.

I do hope that you're only getting positive feedback. . . . And that your comments are taken seriously. As seriously as they deserve to be.

Best,

Nalini Ambady, Ph.D.

Professor, Neubauer Faculty Fellow

Psychology; Tufts University

Dr. Ambady was a social psychologist who did highly influential work on nonverbal behavior and social influences at Harvard, Tufts, and then finally Stanford prior to her death, winning many prizes and honors for her discoveries, but apparently Larry Summers was unable to appreciate their significance.

It is remarkable to me that ten years later, we are repeating a very similar chapter in history after a Google engineer James Damore wrote a company memo detailing why he thought innate differences between men and women explained why there are so few women engineers at Google. His claims again rested heavily on very dubious arguments from evolutionary psychology, just as had Summers's comments, which were in turn largely based on his conversations with Harvard professor Steven Pinker.

What all of these folks—Larry Summers, Steven Pinker, and James Damore, as well as many of the other highly successful white men who have made the same arguments throughout history—strongly believe is that, although more men may be innately better suited for science and engineering than women, there of course should be an individual meritocracy for those women who are as good as or better than men. But what they also entirely fail to see is that individual merit cannot and will not be recognized in the face of pervasive negative stereotyping.

This conclusion is strongly supported by the many studies that show that men are hired over women with equivalent CVs. Moreover, as I reviewed in my essays,[5] negative stereotyping itself is deeply harmful to the ambitions and achievements of women. As Nancy Hopkins ultimately concluded, even when women scientists are highly successful, the research accomplishments of women are perceived as lesser than identical work done by a man.[6]

Everyone should read the final chapter in Malcolm Gladwell's book *Blink* (2005), which describes what happened when major symphony orchestras finally switched to gender-blind auditions. Even then male conductors often persisted in their belief that the women winners were not deserving, and, perhaps most sadly of all, when women won many positions in these orchestras, men no longer saw being a member of these orchestras as prestigious, and salaries dropped.

At present, the evidence that gender-based stereotyping is holding back women's careers is overwhelming, and I am quite tired of hearing unscientifically supported claims from successful white men (unaware of their benefits from their privileged status that continuously fuels their success) that women are innately less able. Given this pervasive negative stereotyping, all of us (male and female) need to be constantly working hard to make the environment more diverse and supportive.

I have focused here on women, but many other groups also face substantial bias and barriers, including Latinos and African-Americans. Despite all good intentions, I am constantly also disturbed by how few Asian people I see in leadership positions at

Stanford and elsewhere. It is very clear looking at the rosters of the National Academy of Sciences that Asian people are only rarely elected no matter how deserving. We all need to do much better!

I am constantly surprised that, given the existence of tenure, more faculty members don't speak up and demand more progress. We need many more people like Nancy Hopkins in this world. Her courageous and long battle to help women at MIT and elsewhere has been a model. There has been a great personal cost to her I am certain. As has been said, leadership is about going out of your comfort zone to help others. No one fits this definition better than Nancy.

I think the reason that many women who reach leadership positions often neglect to use their power to help women is that they may feel that such acts would undermine their leadership authority in the eyes of men. That's why women in leadership positions who do not behave this way are real heroes to me—Jackie Speier (US Congress), Sheryl Sandberg (COO of Facebook), and Drew Faust (president of Harvard University, who eliminated the longstanding and highly sexist Finals Clubs system at Harvard) come to mind.

To paraphrase Martin Luther King, Jr., a first-class scientific enterprise cannot be built upon a foundation of second-class citizens. Change is hard, but we all need to do our part to work toward a better world for all. The welfare of science depends on it, as many studies have shown that diverse perspectives drive innovation. Diverse young scientists frequently are successful because they enter a field and see the same old data in completely new ways.

But it is hard enough to advance the frontiers of science without having to simultaneously confront a mountain of prejudice. Every one of us has the responsibility to work to recognize and lessen these barriers lest the passion for science that drives many of our best and brightest diverse young scientists is extinguished, leading them to "choose" other careers. This tragedy still happens routinely today to women, to LGBT people, to Latinos, African-Americans, and other talented people who are different in some way.

When it comes to prejudice and discrimination, we are all "the monsters." I don't know what it will take to make academia truly welcome to diverse people but I do know that, despite all good intentions and efforts, this work is still only beginning. Despite good intentions, the barriers that diverse talented people continue to experience in academia every day are astonishing. Overall I am happy to say that although many battles are left to be fought, undeniable progress is being made. It was thrilling to visit MIT a few months ago and see that forty years after I had graduated, the faculty finally had a large number of incredibly talented women.

SUMMING UP

As I have described, I believe that my different experiences in life as an LGBT person helped to provide me with diverse perspectives and with the fortitude that I needed to persevere in a competitive world. Growing up transgender in a time of universal ignorance and hate has been difficult and emotionally painful. I believe that most or all of this pain is preventable in a future world where people are less ignorant, more supportive, and more understanding. I have tried my best to help others by being open about my transgender identity and by being as good a scientist, mentor, and human being as I have been able to be. It has been a very great privilege to have had such an enjoyable academic career.

BIOGRAPHICAL NOTES

Born:

Newark, New Jersey

September 13, 1954

Died:

Stanford, California

December 27, 2017

Education:

Massachusetts Institute of Technology, SB (1976)

Dartmouth Medical School, MD (1979)

Harvard Medical School, PhD (1990)

Appointments:

Internship in Internal Medicine, Cornell Cooperating Hospitals (1979–1980)

Residency in Neurology, Cornell Cooperating Hospitals (1980–1983)

Postdoctoral Fellow in Neurobiology, University College London (1990–1993)

Assistant Professor of Neurobiology, Stanford University (1993–1997)

Associate Professor of Neurobiology, Stanford University (1997–2001)

Professor of Neurobiology, Stanford University (2001–2017)

Vice-Chair, Department of Neurobiology, Stanford University (1998–2007)

Chair, Department of Neurobiology, Stanford University (2008–2016)

Director, Masters of Medicine Program for PhD students, Stanford University (2005–2017)

Annexon Biosciences, Inc., Co-Founder and Chairman of Scientific Advisory Board (2011–2017)

Honors and Awards (Selected):

Bell Labs Engineering Scholar (1976)

Diplomate, National Board of Medical Examiners (1980)

Diplomate, American Board of Psychiatry and Neurology (1984)

Mika Salpeter Lifetime Achievement Award, Society for Neuroscience (2008)

Alcon Research Institute Award (2010)

American Association for Advancement of Science (2011)

American Academy of Arts and Sciences (2012)

National Academy of Sciences (2013)

National Academy of Medicine (2014)

Gill Distinguished Scientist Award, University of Indiana (2016)

Ralph Gerard Prize, Society for Neuroscience (2016)

Grundke-Iqbal Award, Alzheimer's Association (2017)

Stanford University President's Award for Excellence through Diversity (2017)

Helis Award for Parkinson's Disease and Neurodegenerative Research (2017)

TRAINEES IN THE BARRES LAB

Trainee Name	Years	Came From	Position in 2017
PhD Students (excluding MD-PhD Students, see below)			
Kaplan, Miriam R.	1994–2000	Brandeis University	Patent agent
Watkins, Trent	2000–2006	University of California, Berkeley	Assistant Professor, Baylor College of Medicine
Rogers, Madalyn	2002–2007	University of Southern Florida	Science writer
Daneman, Richard	2002–2008	McGill University	Assistant Professor of Neuroscience, University of California, San Diego
Winzeler, Alissa	2003–2010	Harvard University	McKinsey—Director of Strategy at Syapse—Precision Medicine
Foo, Lynette	2007–2012	University College London	Scientist, Merck, Switzerland
Scholze, Anja	2009–2014	Pomona College	Scientist, San Jose Science Museum
Guttenplan, Kevin	2013–	Pomona College	Current trainee

(*continued*)

Trainee Name	Years	Came From	Position in 2017
Medical Science Training Program (MD–PhD) Students			
Goldberg, Jeffrey	1995–2002	Yale University	Full Professor and Chair of Ophthalmology, Stanford University School of Medicine
Vargas, Mauricio	2002–2008	University of California, Los Angeles	Ophthalmology Residency, UCLA
Cahoy, John	2003–2007	University of Michigan	Orthopedic Residency, Harvard
Wang, Jack	2009–2014	Stanford University	Neurology Residency, University of California, Los Angeles
Lutz, Amanda	2010–2015	Harvard University	Completing MSTP, about to start residency in neonatology and maternal medicine
Howe, Mariko	2010–2015	Northeastern University	Completing MSTP, about to start residency in pediatric neurology
Sloan, Steven	2012–2016	Johns Hopkins University	Completing MSTP, about to start residency in genetic medicine

(continued)

Trainee Name	Years	Came From	Position in 2017
		Postdoctoral Fellows	
Shi, Jingyi	1994–1997	State University of New York	Washington University, St. Louis
Huaiyu, M. I.	1995–1999	Stanford University	Associate Professor of Preventative Medicine, University of California, Los Angeles
Wang, Songli	1995–2000	University of Pennsylvania	Director of Research at Amgen, South San Francisco
Pfrieger, Frank	1995–1998	University of Constance	Group Leader, CNRS, Neuroscience
Meyer-Franke, Anke	1995–1999	University of Heidelberg	Research Scientist, Gladstone Institute, University of California, San Francisco
Ullian, Erik M.	1998–2003	University of California, San Francisco	Associate Professor of Ophthalmology, University of California, San Francisco
Christopherson, Karen	1999–2004	University of California, San Francisco	Senior Scientist, True North Inc.

(continued)

Trainee Name	Years	Came From	Position in 2017
Harris, Brent	1998–2002	Georgetown	Associate Professor and Director of Neuropathology, Georgetown
Dugas, Jason	1999–2003	University of California, Berkeley	Senior Scientist, Denali Therapeutics
Mandemakers, Wim	2001–2005	University of Rotterdam	Neuroscientist, Dept. of Clinical Genetics, Erasmus MC, Rotterdam
Cayouette, Michel	2002–2004	McGill University	Full IRCM Resident Professor, Full Resident Professor, Department of Medicine, University of Montreal, Adjunct Professor, Anatomy and Cell Biology, McGill University
Stevens, Beth	2004–2008	University of Maryland	Associate Professor of Neurobiology, Harvard Medical School
Eroglu, Cagla	2004–2008	EMBL Heidelberg	Associate Professor of Cell Biology and Neurobiology, Duke University

(*continued*)

Trainee Name	Years	Came From	Position in 2017
Allen, Nicola	2006–2011	University College London	Assistant Professor of Neuroscience, Salk Institute
Zamanian, Jennifer	2005–2010	University of California, San Francisco	Senior Research Associate, Stanford University
Huberman, Andy	2006–2011	University of California, Davis	Associate Professor of Neurobiology, Stanford University
Emery, Ben	2006–2011	University of Melbourne	Associate Professor of Neurology, Oregon Health and Science University/ Jungers Center
Watanabe, Junryo	2007–2011	State University of New York	Teaching Professor, Pomona College
Agalliu, Dritan	2007–2011	Columbia University	Assistant Professor of Neurology, Columbia University
Alexander, Stephan	2009–2013	University of California, San Francisco	Senior Scientist, Merck
Chung, Won-suk	2011–2015	University of California, San Francisco	Assistant Professor of Neuroscience, KAIST, Korea

(continued)

Trainee Name	Years	Came From	Position in 2017
Liddelow, Shane	2012–2017	University of Melbourne	Assistant Professor of Neuroscience, New York University Langone Medical Center
Zhang, Ye	2011–2016	University of California, San Francisco	Assistant Professor of Psychiatry, University of California, Los Angeles
Zuchero, Brad	2011–2016	University of California, San Francisco	Assistant Professor of Neurosurgery, Stanford University
Bohlen, C.	2013–2017	University of California, San Francisco	Senior Scientist, Genentech Inc.
Clarke, L.	2012–	University College London	Current trainee
Fu, Meng-Meng	2013–	University of Pennsylvania	Current trainee
Sun, Lu	2014–	Johns Hopkins University	Current trainee
Bennett, F. Chris	2015–	Stanford University	Current trainee
Li, Tristan	2016–	Duke University	Current trainee

NOTES

LIFE

1. B. A. Barres, L. L. Y. Chun, and D. P. Corey, "Ion Channel Phenotype of White Matter Glia: I. Type 2 Astrocytes and Oligodendrocytes," *Glia* 1 (1988):10–30; B. A. Barres, W. J. Koroshetz, K. J. Swartz, L. L. Y. Chun, and D. P. Corey, "Ion Channel Expression by White Matter Glia: II. The O2A Glial Progenitor Cell," *Neuron* 4 (1990): 507–524; B. A. Barres, W. J. Koroshetz, L. L. Y. Chun, and D. P. Corey, "Ion Channel Expression by White Matter Glia: III. Type 1 Astrocytes," *Neuron* 5 (1990): 527–544.

2. B. A. Barres, L. L. Y. Chun, and D. P. Corey, "Induction of a Calcium Current in Cortical Astrocytes by cAMP and Neurotransmitters," *Journal of Neuroscience* 9 (1989): 3169–3175.

3. B. A. Barres, L. L. Y. Chun, and D. P. Corey, "Glial and Neuronal Forms of the Voltage-Dependent Sodium Channel: Characteristics and Cell-Type Distribution," *Neuron* 2 (1989): 1375–1388.

4. B. A. Barres, L. L. Y. Chun, and D. P. Corey, "Induction of a Calcium Current in Cortical Astrocytes by cAMP and Neurotransmitters," *Journal of Neuroscience* 9 (1989): 3169–3175.

5. B. A. Barres, W. J. Koroshetz, K. J. Swartz, L. L. Y. Chun, and D. P. Corey, "Ion channel expression by white matter glia: II. The O2A Glial Progenitor Cell," *Neuron* 4 (1990): 507–524.

6. B. A. Barres, W. J. Koroshetz, L. L. Y. Chun, and D. P. Corey, "Ion Channel Expression by White Matter Glia: III. Type 1 Astrocytes," *Neuron* 5 (1990): 527–544.

7. B. A. Barres, L. L. Y. Chun, and D. P. Corey, "Ion Channel Phenotype of White Matter Glia: I. Type 2 Astrocytes and Oligodendrocytes," *Glia* 1 (1988): 10–30.

8. B. A. Barres, L. L. Y. Chun, and D. P. Corey, "Ion Channels in Vertebrate Glia," *Annual Review of Neuroscience* 13 (1990): 441–474.

9. B. A. Barres and M. C. Raff, "Proliferation of Oligodendrocyte Precursors Depends on Electrical Activity in Axons," *Nature* 361 (1992): 258–260.

10. B. A. Barres and M. C. Raff, "Proliferation of Oligodendrocyte Precursors Depends on Electrical Activity in Axons," *Nature* 361 (1992): 258–260; B. A. Barres, R. Schmid, M. Sendtner, and M. C. Raff, "Multiple Extracellular Signals Are Required for Long-term Oligodendrocyte Survival," *Development* 118 (1993): 283–295; B. A. Barres, M. C. Raff, F. Gaese, I. Bartke, G. Dechant, and Y. A. Barde, "A Crucial Role for Neurotrophin-3 in Oligodendrocyte Development," *Nature* 367 (1994): 371–375; B. A. Barres, J. Burne, M. Sendtner, H. Thoenen, and M. Raff, "CNTF Controls the Rate of Oligodendrocyte Generation," *Molecular and Cellular Neuroscience* 8 (1996): 146–156.

11. B. A. Barres and M. C. Raff, "Proliferation of Oligodendrocyte Precursors Depends on Electrical Activity in Axons," *Nature* 361 (1992): 258–260.

12. B. A. Barres, M. D. Jacobson, R. Schmid, M. Sendtner, and M. C. Raff, "Does Oligodendrocyte Survival Depend on Axons," *Current Biology* 3 (1993): 489–497.

13. B. A. Barres and M. C. Raff, "Control of Oligodendrocyte Number in the Developing Rat Optic Nerve," *Neuron* 12 (1994): 935–942.

14. B. A. Barres and M. C. Raff, "Proliferation of Oligodendrocyte Precursors Depends on Electrical Activity in Axons," *Nature* 361 (1992): 258–260.

15. B. A. Barres, M. Lazar, and M. C. Raff, "A Novel Role for Thyroid Hormone, Glucocorticoids, and Retinoic Acid in Timing Oligodendrocyte Differentiation," *Development* 120 (1994): 1097–1108.

SCIENCE

1. A. Meyer-Franke, M. Kaplan, F. Pfrieger, and B. A. Barres, "Characterization of the Signaling Interactions That Promote the Survival and Growth of Developing Retinal Ganglion Cells in Culture," *Neuron* 15 (1995): 805–819.

2. A. Meyer-Franke, G. Wilkinson, A. Kruttgen, M. Hu, E. Munro, M. Hanson, L. Reichardt, and B. A. Barres, "Depolarization and cAMP Recruit TrkB to the Plasma Membrane of CNS Neurons," *Neuron* 21 (1998): 681–693.

3. J. L. Goldberg and B. A. Barres, "The Relationship between Neuronal Survival and Regeneration," *Annual Review of Neuroscience* 23 (2000): 579–612.

4. J. Goldberg, J. Espinosa, Y. Xu, N. Davidson, G. Kovacs, and B. A. Barres, "CNS Axon Extension Does Not Occur by Default but Is Stimulated by Electrical Activity Together with Neurotrophic Factors," *Neuron* 33 (2002): 689–702.

5. J. H. Lim, B. K. Stafford, P. L. Nguyen, B. V. Lien, C. Wang, K. Zukor, Z. He, and A. D. Huberman, "Neural Activity Promotes Long-Distance, Target Specific Regeneration of Adult Retinal Axon," *Nature Neuroscience* 19 (2016): 1073–1084.

6. G. Hanson, S. Shen, A. Wiemel, F. A. McMorris, and B. A. Barres, "cAMP Elevation Is Sufficient to Promote the Survival of Spinal Motor Neurons," *Journal of Neuroscience* 18 (1998): 7361–7371.

7. B. A. Barres and B. Stevens, *Purifying and Culturing Neural Cells: A Laboratory Manual* (New York: Cold Spring Harbor Press, 2014).

8. J. Dugas, W. Mandemakers, M. Rogers, A. Ibrahim, R. Daneman, and B. Barres, "A New Purification Method for CNS Projection Neurons Leads to the Identification of Brain Vascular Cells as a Source of Trophic Support for Corticospinal Motor Neurons," *Journal of Neuroscience* 28 (2008): 8294–8305.

9. J. L. Zamanian, L. Zhou, and B. A. Barres, "Astrocytes Promote CNS Neuronal Survival by Secretion of Sulfated Glycosaminoglycans," *Proceedings of the National Academy of Sciences USA* (2017), in preparation.

10. E. M. Ullian, B. T. Harris, A. Wu, and B. A. Barres, "Schwann Cells Strongly Promote Synapse Formation by Spinal Motor Neurons in Culture," *Molecular and Cellular Neuroscience* 25 (2004): 241–251; E. M. Ullian, W. Barkis, S. Chen, J. Diamond, and B. A. Barres (2004) "Invulnerability of Retinal Ganglion Cells to Glutamate Eexcitotoxicity," *Molecular and Cellular Neuroscience* 26:544–557.

11. S. Shen, A. P. Wiemelt, F. A. McMorris, and B. A. Barres, "Retinal Ganglion Cells Lose Trophic Responsiveness after Axotomy," *Neuron* 23 (1999): 285–295.

12. Shen, Wiemelt, McMorris, and Barres, "Retinal Ganglion Cells Lose Trophic Responsiveness after Axotomy."

13. J. Goldberg, M. Vargas, W. Mandemakers, and B. A. Barres, "Inhibition of retinal ganglion cell regeneration by oligodendrocyte derived semaphorin 5A." *Journal of Neuroscience* 24 (2004): 4989–4999.

14. A. Winzeler, W. Mandemakers, M. Sun, M. Stafford, C. B. Phillips, and B. Barres, "The Lipid Sulfatide Is a Novel Myelin-Associated Inhibitor of CNS Axon Outgrowth," *Journal of Neuroscience* 31 (2011): 6481–6492.

15. M. E. Vargas, J. Watanabe, S. J. Singh, W. H. Robinson, and B. A. Barres, "Endogenous Antibodies Promote Rapid Myelin Clearance and Effective Axon Regeneration after Nerve Injury," *Proceedings of the National Academy of Sciences USA* 107 (2010): 11993–11998.

16. A. Brosius-Lutz, W. S. Chung, S. A. Sloan, G. A. Carson, E. Lovelett, S. Posadac, L. Zhou, J. B. Zuchero, and B. A. Barres, "Schwann Cells Use TAM Receptor-Mediated Phagocytosis in Addition to Autophagy to Clear Myelin in a Mouse Model of Nerve Injury," *Proceedings of the National Academy of Sciences USA*, 5 September 2017.

17. J. Goldberg, J. Espinosa, Y. Xu, N. Davidson, G. Kovacs, and B. A. Barres, "CNS Axon Extension Does Not Occur by Default but Is Stimulated by Electrical Activity Together with Neurotrophic Factors," *Neuron* 33 (2002): 689–702; J. Goldberg, R. Daneman, Y. Hua, and B. A. Barres, "An Irreversible, Neonatal Switch from Axonal to Dendritic Growth in the Developing CNS," *Science* 296 (2002): 1860–1864.

18. J. T. Wang, N J. Kunzevitzky, J. C. Dugas, M. Cameron, B. A. Barres, and J. L. Goldberg, "Disease Gene Candidates Revealed by Expression Profiling of Retinal Ganglion Cell development," *Journal of Neuroscience* 27 (2007):8593–8603.

19. A. D. Huberman, M. Manu, S. M. Koch, M. W. Susman, A. B. Lutz, E. M. Ullian, S. A. Baccus, and B. A. Barres, "Architecture and Activity-Mediated Refinement of Axonal Projections from a Mosaic of Genetically Identified Retinal Ganglion Cells," *Neuron* 59 (2008): 425–438; A. Huberman, W. Wei, J. Elstrott, B. Stafford, M. Feller, and B. Barres, "Genetic Identification of an On-Off Direction-Selective Retinal Ganglion Cell Subtype Reveals a Layer-Specific Subcortical Map of Posterior Motion," *Neuron* 62 (2009): 327–334; T. W. Cheng, X. B. Liu, R. L. Faulkner, A. H. Stephan, B. A. Barres, A. D. Huberman, and H. J. Cheng, "Emergence of Lamina-Specific Retinal Ganglion Cell Connectivity by Axon Arbor Retraction and Synapse Elimination," *Journal of Neuroscience* 30 (2010): 16376–16382; J. A. Osterhout, N. Josten,
J. Yamada, F. Pan, S. W. Wu, P. L. Nguyen, G. Panagiotakos, Y. U. Inoue, S. F. Egusa, B. Volgyi, T. Inoue, S. A. Bloomfield, B. A. Barres, D. M. Berson, D. A. Feldheim, and A. D. Huberman, "Cadherin-6 Mediates Axon-Target Matching in a Non Image Forming Visual Circuit," *Neuron* 71 (2011): 632–639.

20. M. Kaplan, A. Meyer-Franke, S. Lambert, V. Bennett, I. D. Duncan, S. R. Levinson, and B. A. Barres, "Soluble Oligodendrocyte-Derived Signals Induce Regularly-Spaced Sodium Channel Clusters along CNS Axons," *Nature* 386 (1997): 724–728; M. Kaplan, M. Cho, L. Isom, R. Levinson, and B. Barres, "Differential Control of Clustering of the Sodium Channels Nav1.2 and Nav1.6 at Developing CNS Nodes of Ranvier," *Neuron* 30 (2001): 105–119.

21. S. Wang, A. Sdrulla, G. diSibio, G. Bush, D. Nofziger, C. Hicks, G. Weinmaster, and B. A. Barres, "Notch Receptor Activation Inhibits Oligodendrocyte Differentiation," *Neuron* 21 (1998): 63–75.

22. J. Dugas, J. B. Ngai, and B. Barres, "Functional Genomic Analysis Revels That Terminal Oligodendrocyte Differentiation Proceeds in Distinct Temporal Stages," *Journal of Neuroscience* 26, no. 43 (2006): 10967–10983.

23. S. Wang and B. A. Barres, "Control of Oligodendrocyte Differentiation by Id2," *Neuron* 29 (2001): 603–614; J. Dugas, and B. Barres, "A Crucial

Role for p57Kip2 in the Intracellular Timer That Controls Oligodendrocyte Differentiation," *Journal of Neuroscience* 27 (2007): 6185–6196; J. Dugas, T. Cuellar, A. Scholze, B. Ason, A. Ibrahim, J. L. Zamanian, L. C. Foo, M. T. McManus, and B. A. Barres, "Dicer1 and miR-219 Are Required for Normal Oligodendrocyte Differentiation and Myelination," *Neuron* 65 (2010): 597–611; J. C. Dugas, A. Ibrahim, and B. A. Barres, "The T3-Induced Gene KLF9 Regulates Oligodendrocyte Differentiation and Myelin Regeneration," *Molecular and Cellular Neuroscience* 50 (2012): 45–57.

24. B. Emery, D. Agalliu, D. Rowitch, and B. Barres, "Identification of Myelin-Gene Regulatory Factor as a Critical Transcriptional Regulator Required for CNS Myelin Gene Expression and Myelination," *Cell* 138 (2009): 172–185.

25. H. Bujalka, M. Koenning, S. Jackson, V. M. Perreau, B. Pope, C. M. Hay, S. Mitew, A. F. Hill, Q. R. Lu, M. Wegner, R. Srinivasan, J. Svaren, M. Willingham, B. A. Barres, and B. Emery, "MYRF Is a Membrane Associated Transcription Factor That Autoproteolytically Cleaves to Directly Activate Myelin Genes," *PLOS Biology* 11, no. 8 (2013): e1001625.

26. A. Meyer-Franke and B. A. Barres, "Astrocyte-Induced Adhesion of Axons and Oligodendrocytes," *Molecular and Cellular Neuroscience* 14 (1999): 385–397; T. A. Watkins, B. Emery, S. Mulinyawe, and B. A. Barres, "Distinct Stages of Myelination Regulated by Gamma Secretase and Astrocytes in a Rapidly Myelinating CNS Co-culture System," *Neuron* 60 (2008): 555–569.

27. Watkins, Emery, Mulinyawe, and Barres, "Distinct Stages of Myelination Regulated by Gamma Secretase and Astrocytes in a Rapidly Myelinating CNS Co-culture System."

28. J. B. Zuchero, M. M. Fu, S. A. Sloan, A. Ibrahim, A. Olson, A. J. Zaremba, J. C. Dugas, S. Wienbar, A. V. Caprariello, C. Kantor, D. Leonoudakus, K. Lariosa-Willingham, G. Kronenberg, K. Gertz, S. H. Soderling, R. H. Miller, and B. A. Barres, "CNS Myelin Wrapping Is Driven by Actin Disassembly," *Developmental Cell* 34 (2015): 152–167.

29. J. D. Cahoy, B. Emery, A. Kaushal, L. C. Foo, J. L. Zamanian, K. S. Christopherson, Y. Xing, J. L. Lubischer, P. A. Krieg, S. A. Krupenko, W. J.

Thompson, and B. A. Barres, "A Transcriptome Database for Astrocytes, Neurons, and Oligodendrocytes: A New Resource for Understanding Brain Development and Function," *Journal of Neuroscience* 28 (2008): 264–278.

30. L. Foo, N. J. Allen, E. A. Bushong, P. B. Ventura, W. S. Chung, L. Zhou, J. D. Cahoy, R. Daneman, H. Zong, M. H. Ellisman, and B. Barres. "A New Method to Purify and Culture Rodent Astrocytes," *Neuron* 71 (2011): 799–811.

31. Y. Zhang, K. Chen, S. Sloan, B. A. Barres, and J. Q. Wu, "An RNA-Sequencing Transcriptome and Splicing Database of Glia, Neurons, and Vascular Cells of the Cerebral Cortex," *Journal of Neuroscience* 34 (2014): 11929–11947.

32. F. Pfrieger and B. A. Barres, "Glial Cells Regulate Synaptic Efficacy," *Science* 277 (1997): 1684–1687.

33. E. Ullian, S. Sapperstein, K. Christopherson, and B. A. Barres, "Control of Synapse Number by Glia," *Science* 291 (2001): 657–661.

34. E. M. Ullian, B. T. Harris, A. Wu, and B. A. Barres, "Schwann Cells Strongly Promote Synapse Formation by Spinal Motor Neurons in Culture," *Molecular and Cellular Neuroscience* 25 (2004): 241–251.

35. K. Christopherson, E. M. Ullian, C. Stokes, C. Mullowny, J. W. Hell, A. Agah, J. Lawler, D. Mosher, P. Bornstein, and B. A. Barres, "Thrombospondins are astrocyte-secreted proteins that promote CNS synaptogenesis," *Cell* 120 (2005): 421–433.

36. C. Eroglu, N. J. Allen, M. W. Susman, N. A. O'Rourke, C. Y. Park, E. Özkan, C. Chakraborty, S. B. Mulinyawe, D. S. Annis, A. D. Huberman, E. M. Green, J. Lawler, R. Dolmetsch, K. C. Garcia, S. J. Smith, Z. D. Luo, A. Rosenthal, D. F. Mosher, and B. A. Barres, "Gabapentin Receptor α2δ-1 Is a Neuronal Thrombospondin Receptor Responsible for Excitatory CNS Synaptogenesis," *Cell* 139, no. 2 (2009): 380–392.

37. H. Kucukdereli, N. J. Allen, A. T. Lee, A. Feng, M. I. Ozlu, L. M. Conatser, C. Chakraborty, G. Workman, M. Weaver, E. H. Sage, B. A.

Barres, and C. Eroglu, "Control of Excitatory CNS Synaptogenesis by Astrocyte-Secreted Proteins Hevin and SPARC," *Proceedings of the National Academy of Sciences USA* 108 (2011): E440–449.

38. N. J. Allen, M. L. Bennett, L. C. Foo, G. X. Wang, C. Chakraborty, S. J. Smith, and B. A. Barres, "Astrocyte-Derived Glypicans 4 and 6 Promote the Formation of Excitatory Synapses Containing GluA1 AMPA Glutamate Receptors," *Nature* 486 (2012): 410–414.

39. B. Stevens, N. J. Allen, L. E. Vazquez, G. R. Howell, K. S. Christopherson, N. Nouri, K. D. Micheva, A. K. Mehalow, A. D. Huberman, B. Stafford, A. Sher, A. M. Litke, J. D. Lambris, S. J. Smith, S. W. John, and B. A. Barres, "The Classical Complement Cascade Mediates Developmental CNS Synapse Elimination," *Cell* 131 (2007): 1164–1178.

40. D. P. Schafer, E. K. Lehrman, A. G. Kautzman, R. Koyama, A. R. Mardinly, R. Yamasaki, R. M. Ransohoff, M. F. Greenberg, B. A. Barres, and B. Stevens, "Microglia Sculpt Postnatal Neural Circuits in an Activity and Complement-Dependent Manner," *Neuron* 74 (2012): 691–705.

41. W. S. Chung, G. Wang, B. Stafford, A. Sher, C. Chakraborty, J. Joung, L. Foo, S. J. Smith, and B. A. Barres, "Astrocytes Mediate Synapse Elimination and Neural Circuit Refinement through the MEGF10 and MERTK Phagocytic Pathways," *Nature* 504 (2013): 394–400.

42. Y. Zhang, S. Sloan, L. Clarke, C. Caneda, C. Plaza, P. Blumenthal, H. Vogel, G. K. Steinberg, M. S. Edwards, G. Li, J. A. Dunctan, S. Cheshier, L. Shuer, E. Chang, G. Grant, M. G. Hayden-Gephart, and B. A. Barres, "Purification and Characterization of Progenitor and Mature Human Astrocytes Reveals Transcriptional and Functional Differences with Mouse," *Neuron* 89 (2016): 37–53.

43. A. M. Pasca, S. A. Sloan, L. E. Clarke, Y. Tian, C. D. Makinson, N. Huber, C. H. Kim, J.-Y. Park, N. A. O'Rourke, K. D. Nguyen, S. J. Smith, J. R. Huguenard, D. H. Geschwind, B. A. Barres, and S. P. Pasca, "Generation of Functional Cortical Neurons and Astrocytes from Human Pluripotent Stem Cells in 3D Cultures," *Nature Methods* 12 (2015): 671–678; S. A. Sloan, S. Darmanis,

N. Huber, T. Khan, F. Birey, C. Caneda, R. Reimer, S. R. Quake, B. Barres, and S. Pasca, "Human Astrocyte Maturation Captured in 3D Cerebral Cortical Spheroids Derived from Pluripotent Stem Cells," *Neuron* 95, no. 4 (2017): 725-727.

44. M. L. Bennet, F. C. Bennett, S. A. Liddelow, B. Ajami, J. L. Zamanian, N. B. Fernhoff, S. B. Mulinyawe, C. J. Bohlen, A. Aykezar, A. Tucker, I. Weissman, E. Chang, G. Li, G. A. Grant, M. Hayden-Gephart, and B. A. Barres, "New Tools for sSudying Microglia in the Mouse and Human CNS," *Proceedings of the National Academy of Sciences USA* 113 (2016): E1738–1746.

45. C. J. Bohlen, F. C. Bennett, A. F. Tucker, H. Y. Collins, S. B. Mulinyawe, and B. A. Barres, "Diverse Requirements for Microglial Survival, Specification, and Function Revealed by Defined-Medium Cultures," *Neuron* 94 (2017): 759–773.

46. R. Daneman, L. Zhou, A. A. Kebede, and B. A. Barres, "Pericytes Are Required for Blood-Brain Barrier Integrity During Embryogenesis," *Nature* 468 (2010): 562–566.

47. R. Daneman, D. Agalliu, L. Zhou, F. Kuhnert, C. J. Kuo, and B. A. Barres, "Wnt Signaling Is Necessary for CNS, but Not Non-CNS, Angiogenesis," *Proceedings of the National Academy of Sciences USA* 106 (2009): 641–647.

48. R. Daneman, L. Zhou, D. Agalliu, J. D. Cahoy, A. Kaushal, and B. A. Barres, "The Mouse Blood-Brain Barrier Transcriptome: A New Resource for Understanding the Development and Function of Brain Endothelial Cells," *PLOS One* 5, no. 10 (2010): e13741.

49. D. Knowland, A. Arac, K. J. Sekiguchi, M. Hsu, S. E. Lutz, J. Perrino, G. K. Steinberg, B. A. Barres, A. Nimmerjahn, and D. Agalliu, "Stepwise Recruitment of Transcellular and Paracellular Pathways Underlies Blood Brain Barrier Breakdown in Stroke," *Neuron* 82 (2014): 603–617.

50. J. L. Zamanian, L. Xu, L. C. Foo, N. Nouri, L. Zhou, R. G. Giffard, and B. A. Barres, "Genomic Analysis of Reactive Astrogliosis," *Journal of Neuroscience* 32 (2012): 6391–6410.

51. S. Liddelow, K. Guttenplan, L. Clarke, F. Bennett, C. Bohlen, L. Schirmer, M. Bennett, A. Munch, W. Chung, T. Peterson, D. Wilton, A. Frouin, B. Napier, N. Panicker, M. Kumar, M. Buckwalter, D. Rowitch, V. Dawson, T. Dawson, B. Stevens, and B. Barres, "Neurotoxic Reactive Astrocytes Are Induced by Activated Microglia," *Nature* 541 (2017): 481–487.

52. A. H, Stephan, D. V. Madison, J. M. Mateos, D. A. Fraser, E. A. Lovelett, L. Coutellier, L. Kim, H. H. Tsai, E. J. Huang, D. H. Rowitch, D. S. Berns, A. J. Tenner, M. Shamloo, and B. A. Barres. "A Dramatic Increase in C1q Protein in the CNS during Normal Brain Aging," *Journal of Neuroscience* 33 (2013): 13460–13474.

53. W. S. Chung, G. Wang, B. Stafford, A. Sher, C. Chakraborty, J. Joung, L. Foo, S. J. Smith, and B. A. Barres, "Astrocytes Mediate Synapse Elimination and Neural CircuitRrefinement through the MEGF10 and MERTK Phagocytic Pathways," *Nature* 504 (2013): 394–400.

54. W. S. Chung, P. B. Verghese, C. Chakraborty, J. Joung, B. T. Hyman, J. D. Ulrich, D. M. Holtzman, and B. A. Barres, "Novel Allele-Dependent Role for APOE in Controlling the Rate of Synapse Pruning by Astrocytes." *Proceedings of the National Academy of Sciences USA* 113 (2016): 10186–10191.

55. Stevens et al., "The Classical Complement Cascade Mediates Developmental CNS Synapse Elimination."

56. G. R. Howell, D. G. Macalinao, G. L. Sousa, M. Walden, I. Soto, S. C. Kneeland, J. M. Barbay, B. L. King, J. K. Marchant, M. Hibbs, B. Stevens, B. A. Barres, A. F. Clark, R. T. Libby, and S. W. M. John, "Molecular clustering identifies complement and endothelin induction as early events in a mouse model of glaucoma," *Journal of Clinical Investigation* 121 (2011): 1429–1444.

57. S. Hong, V. F. Beja-Glasser, B. M. Nfonoyim, A. Frouin, S. Li, S. Ramakrishnan, K. M. Merry, Q. Shi, A. Rosenthal, B. A. Barres, C. A. Lemere, D. J. Selkoe, and B. Stevens, "Complement and Microglia Mediate Early Synapse Loss in Alzheimer Mouse Models," *Science* 352 (2016): 712–716.

58. A. Vukojicic, N. Delestree, E. Fletcher, S. Sankaranarayanan, T. Yednock, B. Barres, and G. Mentis, "Complement and Microglia Mediated Sensory-Motor Synaptic Loss in Spinal Muscular Atrophy," *Society for Neuroscience—Abstracts* (2017).

59. L. Clarke, M. Heiman, C. Chakraborty, and B. A. Barres, "Characterization of the Aging Mouse Astrocyte Transcriptome," *Neuron* (2017), submitted.

60. P. W. Phuan, H. Zhang, N. Asavapanumas, M. Leviten, A. Rosenthal, L. Tradtrantip, and A. S. Verkman, "C1q-Targeted Monoclonal Antibody Prevents Complement-Ddependent Cytotoxicity and Neuropathology in In Vitro and Mouse Models of Neuromyelitis Optica," *Acta Neuropathologica* 125 (2013): 829–840; R. McGonigal, M. E. Cunningham, D. Yao, J. A. Barrie, S. Sankaranarayanan, S. N Fewou, K. Furukawa, T. A. Yednock, and H. J. Willison, "C1q-Targeted Inhibition of the Classical Complement Pathway Prevents Injury in a Novel Mouse Model of Acute Motor Axonal Neuropathy," *Acta Neuropathologica Communications* 4 (2016): 23; Hong, Beja-Glasser, Nfonoyim, Frouin, Li, Ramakrishnan, Merry, Shi, Rosenthal, Barres, Lemere, Selkoe, and Stevens, "Complement and Microglia Mediate Early Synapse Loss in Alzheimer Mouse Models"; Vukojicic, Delestree, Fletcher, Sankaranarayanan, Yednock, Barres, and Mentis, "Complement and Microglia Mediated Sensory-Motor Synaptic Loss in Spinal Muscular Atrophy," *Society for Neuroscience—Abstracts.*

ADVOCACY

1. B. A. Barres, "How to Pick a Graduate Advisor," *Neuron* 80 (2013): 275–279; B. A. Barres, "Stop Blocking the Postdoc's Path to Success," *Nature* 80, no. 2 (2017): 275–279.

2. Barres, "How to Pick a Graduate Advisor"; Barres, "Stop Blocking the Postdoc's Path to Success."

3. B. A. Barres, "Does Gender Matter?" *Nature* 442 (2006): 133–136.

4. Barres, "Does Gender Matter?"

5. Barres, "Does Gender Matter?"; B. A. Barres, "Neuro Nonsense. Book review of *Delusions of Gender* by Cordelia Fine," *PLOS Biology* 8 (2010): e1001005.

6. Courtney Humphries, "Measuring Up," *MIT Technology Review* 120 (August 16, 2017)

INDEX

INDEX